THE
CALIFORNIA SKY WATCHER

H
HEYDAY
1974
2024
50

THE CALIFORNIA SKY WATCHER

UNDERSTANDING WEATHER PATTERNS AND WHAT COMES NEXT

WILLIAM A. SELBY

Berkeley, California

Library of Congress Cataloging-in-Publication Data

Names: Selby, William A., author.
Title: The California sky watcher : understanding weather patterns and what comes next / William A. Selby.
Description: Berkeley, California : Heyday, [2024]
Identifiers: LCCN 2023049902 (print) | LCCN 2023049903 (ebook) | ISBN 9781597146401 (paperback) | ISBN 9781597146432 (epub)
Subjects: LCSH: Meteorology--California--Popular works. | Climatic changes--California--Popular works. | Atmospheric pressure--California--Popular works.
Classification: LCC QC984.C2 S45 2024 (print) | LCC QC984.C2 (ebook) | DDC 551.509794--dc23/eng/20231220
LC record available at https://lccn.loc.gov/2023049902
LC ebook record available at https://lccn.loc.gov/2023049903

Cover Images (from top to bottom): Galyna Andrushko (Sierra Nevada), Joseph Sohm (Antelope Valley), Scalia Media (Salton Sea), Scalia Media (Joshua Tree National Park). All images licensed from Shutterstock.com
Cover Design: *the*BookDesigners
Interior Design/Typesetting: *the*BookDesigners

Published by Heyday
P.O. Box 9145, Berkeley, California 94709
(510) 549-3564
heydaybooks.com

Printed in China by Regent Publishing Services, Hong Kong

10 9 8 7 6 5 4 3 2 1

CONTENTS

INTRODUCTION

CONNECTING TO OUR ATMOSPHERE

What is your relationship with the weather? I am and have always been a weather nut. I am also a native Californian born in Long Beach. My brother and I grew up on the western edge of Santa Ana in a tiny one-and-a-half-bedroom house that made a little footprint on two open lots. In what must have been one of the most diverse neighborhoods in the country at the time, almost everyone spent most of their days outside after completing the few necessary indoor tasks. Our house was small, with a claustrophobic feel to it. There was no reason to be inside when outdoor sports, garden chores, and other activities constantly called us out. Our iconic California weather almost always cooperated.

I remember the extreme openness of those early years when we could walk to the end of our street and enjoy views of what seemed to be endless rows of field crops across the flood-plain. Everyone on our entire block drank and irrigated from a community well that pumped water to all of our houses and was operated and maintained by all of our neighbors. We didn't even have water meters; seemingly limitless water was there for the taking. Years later, as the region's open spaces filled with developments, the City of Santa Ana annexed our neighborhood. City amenities and services followed. We considered them luxuries: a freshly paved and maintained street with sidewalks and street lights and even a sewer line to replace our septic tank. Although our street name changed during this period of accelerated development, there remained a sense of openness, where natural light spread out to the horizon. It was impossible to ignore the vast sky and changing weather patterns that played such important roles in everyday life.

Perhaps this explains why—before I was ten years old—I realized how my fascination with the weather was more powerful than for the other kids . . . and quite unique. Why wasn't everyone so interested in the weather? Even in such an environment where people were constantly obliged to look up at the sky and evaluate changes in the atmosphere, I was far more obsessed and consumed with learning about those changes. Like most other excessively conscientious kids hoping to fit in, I often tried to keep this infatuation to myself. Nobody wants a reputation as the weird kid on the block. So my sacred love affair with the sky became a deeply personal way to find peace and solitude separate from life's problems. I was fortunate to remain curious and determined to pursue my curiosity as I accumulated and analyzed every day's and night's weather observations, gathering as much information from as many sources as possible to learn what was going on up there. I even made my own forecasts to try to outwit professional weather forecasters. Not so much luck there.

This obsession emboldened my search to unravel the many mysteries of our atmosphere and our world. It helped lead me through school and eventually pushed me into Northern California to complete my graduate degree and to work in a more changeable climate. My expanding love of science and nature led me into a rewarding career as a science teacher and then professor of geography and earth science. My fondness for the Golden State fueled explorations from north to south and from the coast to the mountains and the deserts, researching, living in, and appreciating every corner of California.

And so I must thank our tiny house, which pushed me out into our big lot in our working-class neighborhood. There, I discovered the magic of science and weather that has colored my life and brought me so much entertainment and satisfaction. Even today, when so many of us nearly forty million Californians are surrounded by crowded urban chaos, we each have the sky, a common and shared liberation. Its changeable energies can nurture us and destroy us. Its patterns and seasons shape our bodies and minds. The sky, a source of wonder and awe for our entire existence as thinking creatures, can soothe any nature deficit disorders when we lift our eyes upward.

I've been watching the sky and studying its science for more than five decades. My stories might help remind you that no matter where you find your magic in this world and in this life, you should follow it. You can start here by following me on this perpetual journey to explore the incredible diversity of questions and the unlimited learning and life experiences found in our California skies, in the weather and climate of California.

There can be beauty and magic or tragedy and heartbreak in every day and experience. But the real miracle is that we are here to appreciate the clouds or blue sky and to feel the breeze and to wonder how it all fits together. A never-ending series of connected and ever-changing weather events and seasonal cycles surrounds and follows us from birth until death, ushering us through this life. Attending to these cycles and their impacts can also help us better appreciate how our weather connects all of us and plays such a vital role in everything we do, everything we are, and everything we wish to be.

This book offers the most convenient, least expensive, and lowest impact ecotourism and natural science learning opportunities you will ever experience. It doesn't require you to sign up or to endure the formalities and hassles of a traditional class. It doesn't require any entrance or parking fees or reservations at a park, campground, hotel, or B&B. You don't have to jam yourself into a plane, train, or shuttle or put gas in your car. You don't even have to pack your bags or plan for unexpected or unfortunate events that could ruin your vacation or interrupt your field excursion. Our accessible sky dome is waiting for your attention: gaze outside or walk out your door. Our extreme science laboratory is running experiments that you can observe and experience by simply sensing your air and sky.

And everyone, no matter who they are or how much money they earn, can admire the same sky and sense the same weather. Many miles away, people are looking straight up to see that very cloud you are viewing near the horizon. They could be idealistic humanitarians or callous crooks; like it or not, our sky dome conjoins us. Our egalitarian weather is constantly calling out to all of us, connecting us and reconnecting us to our source. So I challenge you to reopen your heart and your mind to views and dimensions that offer fresh perspectives and new opportunities to appreciate and celebrate this life on this planet.

FIGURE I-1. Crepuscular rays peek through a mix of summer storm clouds building over the mountains during a rare invasion of tropical weather. Photo by Matt Wright.

Anyone enthusiastic about atmospheric science would be excited to spend time in California, so I count myself lucky to have observed the enormous diversity of California's weather and climates for my whole life. California contains examples of most major climates on Earth, except for the tropics—there are no humid equatorial rainforests here. Each year, somewhere and someone in the state is at least temporarily impacted by nearly every weather event and air mass that can be experienced on Earth. Just as you can enjoy a cultural and culinary tour of the world by traversing California, the same goes for our state's astounding assortment of landscapes, weather patterns, and climates. These weather stories are learning opportunities that can be applied wherever you live or travel, even if you are not chasing or fleeing that evanescent California Dream.

Look out your window. What kind of weather are you experiencing right now? How can you better understand these observations and predict what might happen next? These are questions all of us ask. Our curiosity is piqued during thunderstorms, heat waves, cold snaps, shifting winds, and other unusual weather events. Nature's shows invite us into the mysterious and fascinating stories that make life on Earth more rewarding and exciting, stories that teach us about where we came from, why we are here, and where we are going. Countless philosophers, naturalists, and scientists have attempted to put these stories into words. The best encourage us to look through a window rather than into a mirror. The window allows us to see and experience reality outside ourselves.

Are the idyllic carefree California atmospheric stereotypes true, or are they mere meteorological myths championed by chambers of commerce, the tourist industry, and real estate interests? The correct answer depends on the day and your location, since you will encounter strikingly different weather conditions in each region of the Golden State and surprisingly dramatic changes over time. Temperatures soaring over 110°F are suddenly interrupted by violent flash floods generated by severe thunderstorms. Several feet of snow accumulate in wind gusts over 100 mph as temperatures plummet below 0°F. Relentless Pacific storms drive cool, drenching rains to saturate soils, soak forests, and engorge streams and rivers day after day. Name the place and time, and you can find just about any weather forecast and extreme in the Golden State. Such atmospheric restlessness implores us to explore California's astonishingly diverse and constantly changing weather patterns and climates.

So keep asking yourself why that beautiful billowing cumulus cloud is growing and be curious about the processes that are shaping it. Ask how those cirrus clouds streaming above 25,000 feet produce such brilliantly red sunsets, colorful halos, and sundogs. What caused the latest drought or flood? The answers may range far beyond a simple understanding of our weather as they lead you to natural wonders that will help you understand our world and your place in it.

Before launching into our seasonal tours across California, we should pause to cover some fundamental concepts that will make your weather and climate experiences easier and a lot more fun. For us, the sky is our classroom, and we'll start our seminar with the most powerful weather influencer of all: air pressure. To understand the weather means spending some time with the fundamental physics of atmospheric science. If you still feel the overwhelming need to ditch class for now and skip right into our seasonal weather expeditions, you know that the following pressure and wind clarifications will be waiting for you when you need them.

HOW AIR PRESSURE COMMANDS OUR WINDS AND WEATHER

Have you felt under pressure lately? That might be because there are about 14.7 pounds per square inch of air pressure impacting every surface (including your body) near sea level, which will force about 29.92 inches of dense mercury up into a narrow vacuum tube that we call a barometer. This pressure is exerted by many billions of highly energetic air molecules colliding against one another (and every surface facing every direction) roughly 10 billion times each second. Meteorologists measure this average sea-level air pressure as one atmosphere, or 1,013 mb (millibars), or 1013 hPa (hectopascals), the units of barometric pressure used in forecasts and weather maps. But they are all simply measuring the forces exerted by these bombarding molecules. These billions of air molecules might be invisible, but precise weather predictions rely on precise air-pressure measurements because high- and low-pressure systems command major weather patterns and conditions in California.

Air pressure changes as air density and temperature fluctuate. Increasing either air temperature *or* air density will result in increasing pressure within a closed container, since the volume remains the same. As a mini experiment to demonstrate how air pressure changes, place an empty jar or other container in the refrigerator and then seal it. Later, take the sealed container out of the refrigerator and place it in your warm kitchen for a while. When you finally take the seal off, you will notice that the higher temperature in the container has increased the pressure as highly energetic molecules collide, causing some air to pop out. (If your jar also contains a liquid food product, such as salad dressing, ketchup, etc., you might end up with a small but messy explosion in your kitchen.) That's high pressure. Now, instead of changing the temperature, pump air into a sealed container to increase the density (such as when pumping up a tire or soccer ball) and you will also notice the increase in air pressure. But the oversimplified gas law (pressure = temperature × density) won't get you far in predicting the weather, because our atmosphere is not a closed container. Instead, we are surrounded by a highly dynamic open-air environment, where air parcels (pockets of air) are free to expand and contract in volume. This allows temperature and density to react and readjust to each other in a more complicated dance that determines air pressure.

FIGURE I-2. This bag of chips and air was sealed near sea level. We are now in the White Mountains, just over 10,000 feet above sea level, where outside air pressure is about 30 percent lower. High-pressure air inside the closed bag attempts to expand and push into the outside high-altitude, low-pressure atmosphere. By contrast, open-air parcels in the background are liberated to expand as they rise up into higher altitudes; they cool to their dew points, forming puffs of cumulus clouds above the bristlecone pines.

The open atmosphere is constantly being pulled down and compressed toward Earth's surface by gravity, but is also charged with energy. Gravity keeps molecules in the atmosphere from flying out into space, but energy keeps molecules gyrating around and colliding, causing air columns to expand far above the surface (figure I-3). Colder, less energetic air masses are denser, so the chilly air at the North and South Poles is heavy, and hugs the Earth's surface. Meanwhile, near the equator, warmer and highly energetic air masses are expanding and rising, pushing air columns higher. As they move, tall stacks of expanding warmer air migrate north toward California from the equator, while very dense, compact, shorter stacks of cold air may drift south toward California from colder regions. When air is warmed (as when heated by land surfaces during the day), it may expand in volume so drastically that air density decreases much more than temperatures first increased; this often results in low atmospheric pressure near Earth's heat-radiating surfaces. (Refer back to our simplified gas law: pressure = temperature × density.) So low-pressure air masses, known as thermal lows, commonly form above desert surfaces during hot summer afternoons. By contrast, air masses that form over the cold land during winter become dense and compact as slower-moving air molecules pack together. This increase in density near the surface overcomes the initial consequences of decreased air temperatures (our simplified gas law again). The Great Basin High pressure that often settles over cold western states during autumn and winter is a classic example.

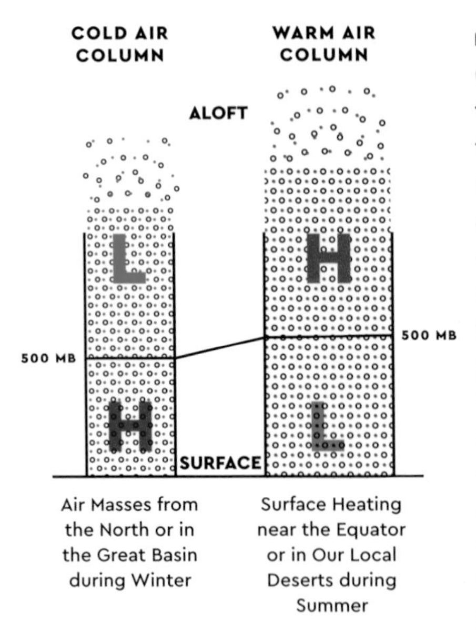

FIGURE 1-3. Cold air columns (such as near the poles and the Great Basin during winter) and warm air columns (such as near the equator and over California deserts during summer) display very different pressure characteristics. This idealized depiction illustrates how the 500 mb level (approximately halfway through the vertical mass of air columns) rises to higher altitudes in the warmer expanding air and is pulled down to lower altitudes in the colder, more compact air. This gets more complicated (and exciting) in California and other middle latitudes, where contrasting air masses often migrate, mix, and do battle.

All of this shifting pressure, density, and temperature is a big weather deal because less dense, more buoyant air masses with low atmospheric pressure will want to rise (the weather definition of *unstable*), possibly causing stormy weather. By contrast, heavy, dense air masses with higher air pressure tend to sink and settle (defining stable air), resulting in fair weather. As solar radiation warms Earth's surface, the heated air above may expand and become unstable, especially when nature has some moisture to work with. This makes thunderstorms more common in the afternoons (compared to cooler mornings) during summer monsoon season in the Desert Southwest. During winter, when very cold, dense air masses swoop down from the north and west, they lift up relatively warm, moist, less dense air masses along the way; unstable, rising low pressure that forms along these turbulent frontal boundaries creates much of the storminess Californians know as the winter rainy season. Cold, dense, high-pressure air often settles behind these storms. Meanwhile, because air pressure always seeks equilibrium (think of the way air rushes out of a deflating balloon), differences in these areas of high and low pressure set air in motion, pushing winds out of high-pressure systems and pulling winds into low-pressure systems.

While warm or cold land and water surfaces shape some of our weather, upper-level pressure and winds are often the big weather makers around California latitudes. When upper-level winds begin to slow down and pack closer together (high up in the atmosphere above about 18,000 feet [5,600 m]), the air up there gets denser; such heavy air will want to sink toward the surface. When warm, rising air near the equator spreads out miles high and flows toward California, the winds are forced to pile up (converge) high above our heads. By the time these upper-level winds reach near 30 degrees latitude (just south of California), they slow down and cram together to become heavy, dense high-pressure systems.

These high-pressure systems (known as subtropical highs) shift a few or several degrees north during summer. Because they contain such dense, stable (sinking) air, they represent stubborn blockades to storms that might otherwise invade the Golden State. Welcome to our long summer droughts—shallow, gray marine layers (cool, moist, shallow layers of air that form over the ocean) may get trapped below the descending air, but storms are rare. As winter approaches, our subtropical high-pressure systems shift south with the sun, opening the door to migrating unstable (rising) air masses and low-pressure storms from the North Pacific Ocean. Since Northern California latitudes are farther from the dense, stable, high-pressure fair-weather air masses, they normally experience longer and wetter winter rainy seasons. Southern California latitudes remain closer to these great protectors (sometimes characterized as resilient ridges) and therefore usually experience longer summer droughts and lower annual precipitation.

As cold air masses migrate from near the Arctic toward the south during winter, they also push major boundaries between relatively colder northern air and warmer middle-latitude and subtropical air farther south toward California. This creates strong temperature and pressure gradients (the difference between high and low pressure over a distance), which propel high-velocity upper-level winds called jet streams. The polar jet stream meanders as it generally flows from west to east around our middle latitudes (30–60 degrees). When those cold air masses (dense and compact) spread south in winter, they push temperature and pressure boundaries—and an energized upper-level polar-front jet stream—south ahead of them. Giant bends in wind-flow patterns (called Rossby waves) form that steer the jet stream south, similar to a river that meanders. These baggy loops that dip south (called upper-level troughs) often bring unstable weather and precipitation south into California. By contrast, when relatively warmer air masses (with taller air columns) migrate northward, unstable boundaries are pushed far north of our state, and we experience fair weather. This usually occurs during summer when the jet stream is directed north around giant omega-shaped Rossby waves, high-pressure ridges that produce fair weather.

In California, our location in relation to these meandering upper-level wind patterns (usually flowing and looping from west to east within those Rossby waves) often drives our weather (figure I-4, p. 15). How does this work? Let's simplify the science here. These high-level, high-velocity winds are balanced by two forces, the pressure gradient and the Coriolis effect (the force created as the Earth turns beneath the wind). The pressure gradient always pushes wind out of high pressure and pulls it into low pressure. But because the Earth turns under the wind, the Coriolis force is always pulling the wind toward its right (not your right) in the Northern Hemisphere. Thus a balance between these two forces sends these upper-level winds flowing along, meandering up and over high-pressure ridges and down and around low-pressure troughs.

As the upper-level wind makes its turn up and over a ridge, it slows down and bunches together (converges) aloft, in a manner similar to what happens to freeway traffic when it encounters a couch accidentally dropped in the lanes ahead. This increases the density of the air so that it becomes heavier and sinks toward the surface. Also note how winds are encouraged to spin clockwise while rounding the upper-level ridge. We are now in a region where air masses are likely to be sinking toward the surface and spinning clockwise, which is why we call them anticyclones. Stable, fair weather is the result below and just downwind (just east) of upper-level high-pressure ridges. By contrast, when these upper-level winds make their turn down and around the baggy troughs, they are free to spread out (diverge), creating a void aloft, similar to traffic after we all thread past that couch that fell off a truck in the fast lane. We are liberated to hit the accelerator. These diverging and accelerating winds aloft may encourage air parcels to rise from the surface, resulting in unstable stormy weather below and just downwind (just east) of upper-level troughs. Further, note how the wind is encouraged to spin counterclockwise as it turns around the trough and back toward the north. As air masses are rising and being stretched from the surface, their counterclockwise spin will accelerate, similar to how ice skaters spin faster as they stretch higher. So, regions below and downwind of upper-level troughs often experience cyclogenesis—meaning that storms (cyclones) are born and strengthen—especially during our winter months. This contrasts with the heavy, dense, stable, high-pressure air masses that are sinking, flattening, spreading, and spinning clockwise toward the surface, so common during our summer months.

All these atmospheric players dancing, diverting, blocking, and blowing create our seasonal weather patterns. During summer, the North Pacific Subtropical High pressure ridge of stacked air (described further in chapter 1) will dominate our weather. When it strengthens and/or moves closer to us, stable air masses are heated by compression as they descend over us. Temperatures rise (especially inland), and marine layers are squashed closer to the surface or driven out to sea. When the high-pressure ridge weakens or moves away, marine layers thicken and encroach farther inland with the stronger onshore flow (sea breeze), resulting in cooler summer days. During winter, after the protective shield of the North Pacific Subtropical High has shifted farther south,

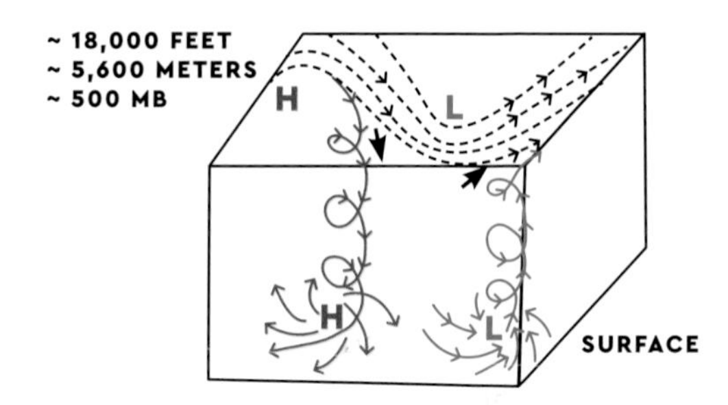

FIGURE I-4. Powerful connections are evident between upper- and lower-level pressure and winds. These upper-level meandering troughs and ridges and their winds and jet streams loop and migrate over middle latitudes across the globe, escorting fair and stormy weather near the surface.

upper-level troughs drive southward, pushing ascending, unstable low-pressure air masses and storms along their way, delivering our turbulent winter rainy seasons.

A complex and interconnected series of systems and cycles is constantly circulating air out of high-pressure systems and into low-pressure systems around us. The moving air molecules you sense are mere sails in the winds powered by pressure gradients that will continue to carry them many thousands of miles distant and thousands of vertical feet in elevation. And although surface heating and cooling do affect and even produce many of our wind and weather patterns, those upper-level pressure patterns and winds drive most major changes in California weather. Interactions among all the constantly flowing and cycling upper- and lower-altitude air masses, pressure systems, winds, and weather patterns challenge us to appreciate our dynamic atmosphere.

Now we can better understand the forces that control the speed and direction of winds at Earth's surface. It all starts with the pressure gradient force that sets air in motion from high- to low-pressure areas. When there is little or no difference between high and low pressure at your location in relation to your surroundings, the air is calm. The greatest differences between high- and low-pressure systems in closest proximity produce the strongest winds. For instance, when dense, clear air returns following deep low-pressure storms across California (typical of late fall, winter, and spring), high winds may sweep through, especially after frontal boundaries have passed by. But during late spring and summer, strong onshore winds attempting to fill the space left by rising hot air over the hot deserts will carry cool air inland, always attempting to establish equilibrium. These winds often reverse in autumn and winter when the heavy, dense continental air masses over cold land drain out to fill in the lighter, warmer air over the ocean, creating offshore winds, especially during colder nights and early mornings.

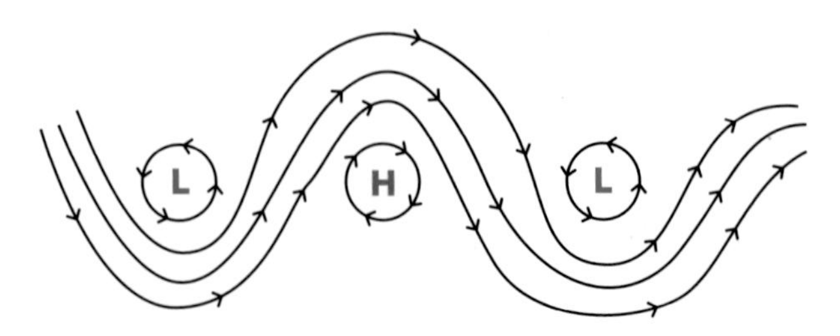

FIGURE 1-5. This idealized upper-level map (at around 500 mb or 18,400 feet [5,600 m]) demonstrates pressure and wind trends at least halfway (in mass) up through our atmosphere. Notice the undulating Rossby waves that circle the planet. An upper-level balance is achieved between the pressure gradient force and the Coriolis force, resulting in winds meandering around Earth roughly parallel to the isobars (imaginary lines connecting points of equal pressure).

The wind's speed and direction in the upper atmosphere is ruled by Earth's spin (Coriolis effect) beneath these huge pressure systems and gradients miles above us. But at the immediate surface, as the wind drags on rough seas or bumpier land topography, this friction slows it down and makes it more subject to local pressure differences. Surface friction slows the wind and weakens the Coriolis force just enough to allow the pressure gradient force to win in their tug-of-war so that the wind

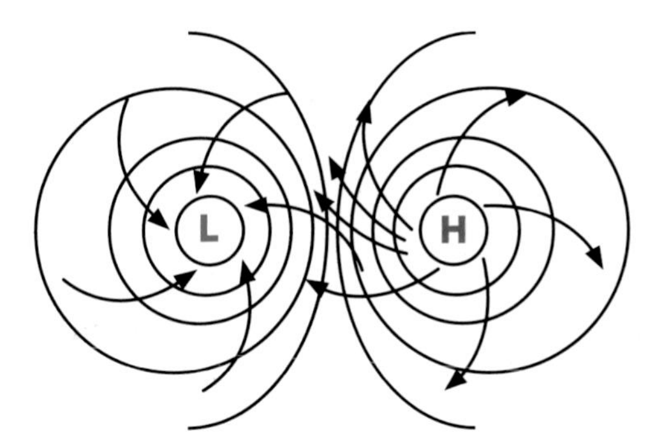

FIGURE I-6. When friction is introduced near the surface, the pressure-gradient force overcomes a weakened Coriolis force, causing winds to cross isobars and turn clockwise *out* of high-pressure systems and counterclockwise *into* low-pressure systems. The strongest winds will blow across the steepest (most extreme) pressure gradients. Your location in relation to these fluctuating and migrating highs and lows will determine the strength and direction of winds you experience at the surface.

can blow *across* isobars (imaginary lines connecting points of equal pressure). Consequently, winds pour *out* of high-pressure systems (though still curving clockwise) and *into* low-pressure systems (though still curving counterclockwise) in the Northern Hemisphere. Because cyclones are spinning counterclockwise, relatively warm winds from the south often blow into storms approaching from our northwest, then veer to bring colder air out of the west or north after the storms pass by. By contrast, that clockwise spinning high pressure over the Great Basin during autumn and winter may push surface winds from the northeast out over California and toward the ocean.

These basic concepts should help you better understand specific weather patterns considered throughout this book. OK—pressure and wind class is dismissed. Now that you better understand how the atmospheric wind engine works, we can jump onto our wild weather rides!

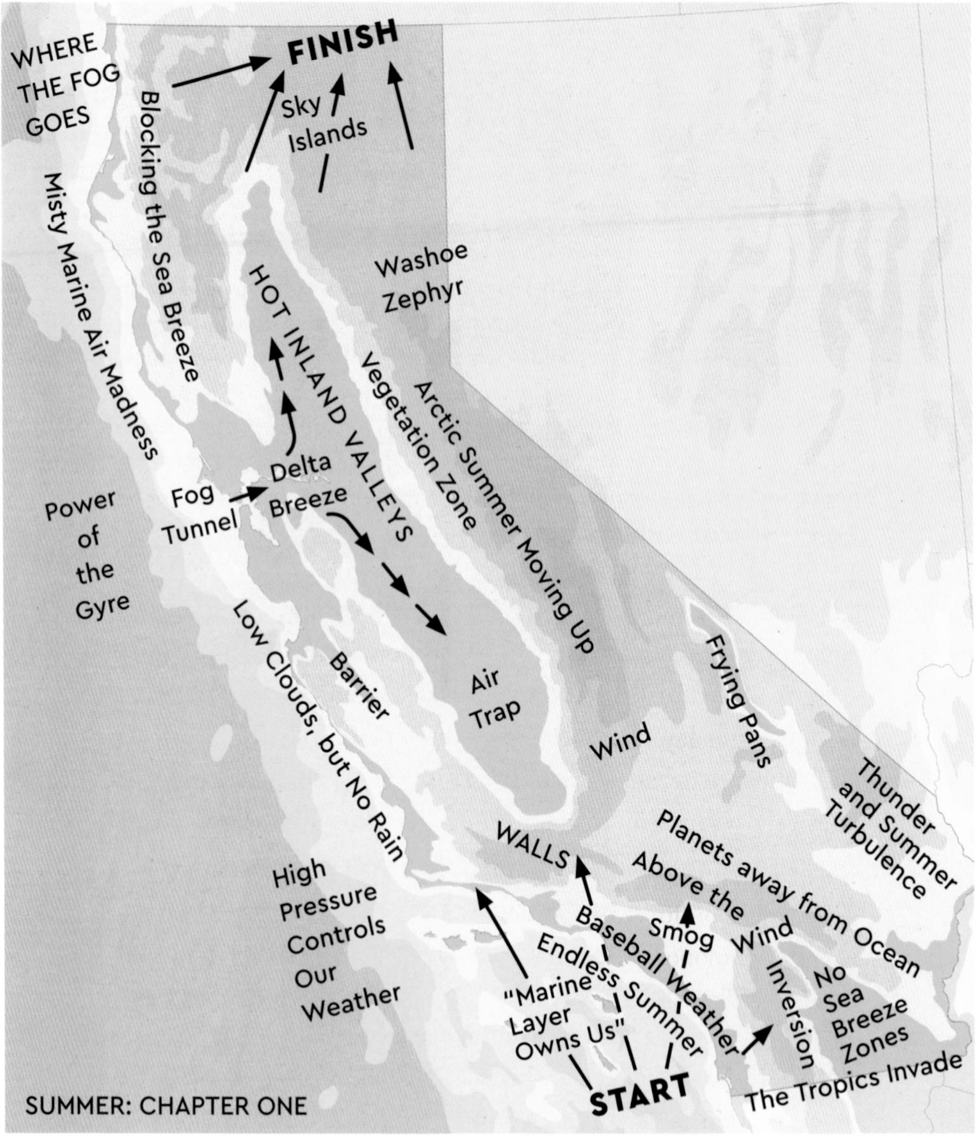

WHERE THE FOG GOES

FINISH

Sky Islands

Blocking the Sea Breeze

Misty Marine Air Madness

Washoe Zephyr

HOT INLAND VALLEYS

Vegetation Zone

Arctic Summer Moving Up

Fog Tunnel

Delta Breeze

Power of the Gyre

Low Clouds, but No Rain

Barrier

Air Trap

Frying Pans

Wind

Thunder and Summer Turbulence

Planets away from Ocean

WALLS

Above the Wind

Smog

Baseball Weather

Endless Summer

High Pressure Controls Our Weather

"Marine Layer Owns Us"

Inversion

No Sea Breeze Zones

The Tropics Invade

START

SUMMER: CHAPTER ONE

CHAPTER ONE

SUMMER'S STABLE DROUGHT

Late night and early morning low clouds and fog, otherwise mostly sunny during the afternoons. High temperatures will range from the 70s at the beaches to the 80s on the coastal plain, dropping into the low 60s overnight. Mainly calm winds except for gentle afternoon sea breezes. No rain and little change in the long-range forecast.

The next day: ditto. The day after that: same again.

Millions of Californians are familiar with the ideally comfortable coastal California weather forecasting monotony that plays out day after day for months, especially from every May into September. Our trouble-free atmospheric mildness has been featured in art, poems, essays, books, music, and movies since visitors and settlers wandered and then poured into California. It resonates with those who celebrate it and is ridiculed by those who are energized by the erratic turbulence of weather extremes.

Our weather journey begins with these archetypal mild summer climates along the Southern California coast. As this chapter progresses, we will eventually travel north and inland across deep valleys and over lofty mountains to experience California's more extreme and changeable climates, until we end our summer journey in the far northeast corner of the state.

Our iconic summer weather is revered from Santa Barbara (called the American Riviera by some) south into Baja. Along these coastal strips, you will find some of our planet's mildest climates, which frequently live up to their stereotypes. Some of the larger populations of Native Americans enjoyed this comfortable near perfection for thousands of years before the great invasions and the evolution of California's renowned beach and surf cultures. Our paradisiacal weather was shamelessly exploited in a twentieth-century culture increasingly dominated by advertising and image. It became a trope so ubiquitous, it was mocked into parody.

Those who have been fortunate enough to live along Southern California's coast sense more than a degree of truth in the praise and the parody. After all, this was the weather that attracted folks ranging from plein air (open air) artists from around the world searching for brilliant light, to patients recovering from respiratory diseases. It was the environment that beckoned people who were escaping harsher climates in the US and beyond. Southern California coastal climates have always ranked at the top of the list of reasons why millions of people and a host of industries, ranging from agriculture to aerospace to technology to film and other entertainment, flocked to these coastlines.

The growth brought crowds that eventually jammed the roads and belched air pollution as the state's population swelled toward forty million and its economy grew to become the fifth most powerful in the world. Bustling masses began squeezing the farms and renowned citrus groves out to more distant, less comfortable, and less desirable inland valleys. Consequently, these same mildest, sunny coastal strips are now some of the most crowded and expensive communities in the country. More recently, average wage earners have also been bumped out to those distant and more affordable inland locations, where seasons are harsher and the spent remains of exhausted sea breezes often dump the smog. This is also one reason why California has made great efforts to keep its beaches public and accessible to anyone who wants to sacrifice the travel time and make the effort to visit them.

FIGURE 1-1. The setting sun's enduring longer wavelengths of red and orange briefly illuminate the bottoms of high, ice-crystal cirrus clouds. Millions of sightseers have witnessed similar spectacles from California beaches.

WHEN HIGH PRESSURE CONTROLS OUR WEATHER

Steadfast high pressure sums up the weather throughout most of California during summer. Such dense, heavy air sinks out of high-pressure systems, creating stable, fair weather. By contrast, because less dense air is lighter, it rises up within low-pressure systems. Such unstable air may expand and cool to its dew point, often resulting in stormy weather. (Dew point is reached when the air cools to reach its saturation temperature—100 percent relative humidity.) Because it spreads out of the highs and drains into the lows, wind always tends to blow from relative high-to-low pressure areas at the same altitude.

One of the most massive pressure systems in the world builds and then anchors itself over the northeastern Pacific Ocean, especially from spring through September. Dense, heavy air descends out of this resilient ridge of high pressure, blocking transient low-pressure systems from intruding into California. This North (or East) Pacific Subtropical High is California's great protector and seemingly perpetual producer of stable air masses. It commands summer's drought and often determines the extent of cooling sea breezes and severity of heat waves as it wobbles, shifts around, expands, and contracts farther out to sea one week and then nudges over our state the next. It sometimes earned the notorious "ridiculously resilient ridge" moniker (a term coined by climate scientist Daniel Swain) during our severe droughts. (In the most general terms, California has two types of droughts. Our Mediterranean climate is defined by each year's regular summer-season drought. But when precipitation totals fall below the long-term average for a series of seasons or years, longer anomalous and exceptional droughts may develop.) As surface winds flow out from the high's center, toward relatively lower pressure, they are veered to their right in the Northern Hemisphere by the Coriolis effect. The result is the North Pacific Subtropical High, a gigantic clockwise pinwheel of wind that dominates over the North Pacific Ocean. These surface winds turn toward the south as they drain out of the eastern side of the High (our West Coast). Cyclones (storms) spin winds in the same direction as Earth's rotation, and anticylones spin in the opposite direction of Earth's rotation; the North Pacific High is an anticyclone. And it is these clockwise anticyclonic winds that drive the California Current, that wide, cold stream of ocean water drifting from north to south along our coast. (During late summer, a very different subtropical "Four Corners High" may nudge in from the east, sometimes bringing moist summer heat.)

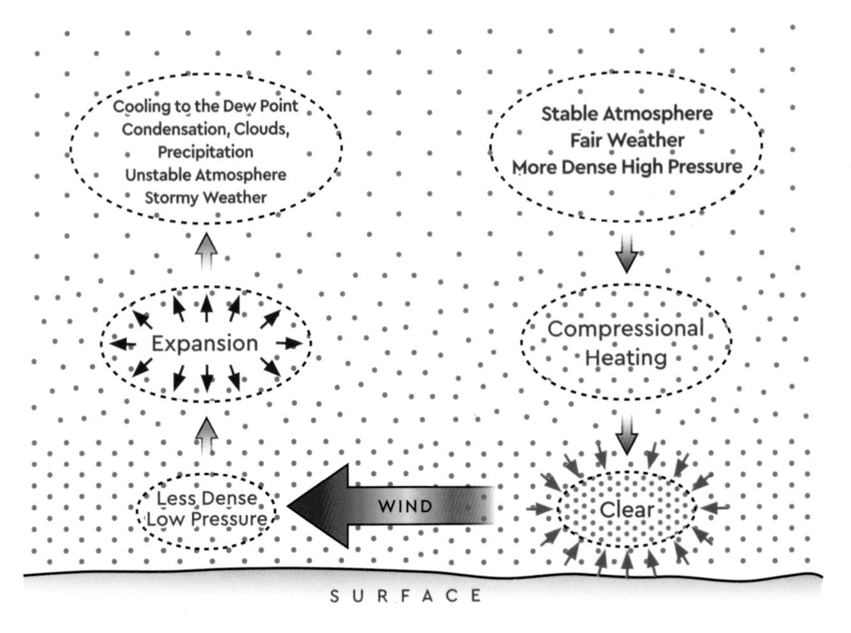

FIGURE 1-2. Air tends to sink and circulate out of high-pressure areas and then flow into low-pressure areas.

As inland surfaces are warmed, the Earth transfers heat upward, causing air to expand, creating thermal low pressure near the surface. Summer winds flow off the cool ocean and toward these heated inland basins. As they are sucked onshore toward this inland void, the winds may be blocked by coastal mountains, siphoned into coastal valleys, or funneled through narrow canyons. These cool, heavy, dense air masses in coastal California are among the most stable on Earth. Resulting marine layers, often carrying the region's famous coastal low clouds and fog (often called advection fog because it forms over the cold ocean and drifts inland), are trapped below summer's high pressure. Stop to sense these feathery-soft summer sea breezes as they drift within marine layers that have settled and pooled in coastal valleys. You are feeling the molecules that make up the air itself as they rush by you, pouring out of high-pressure areas and draining into low-pressure areas.

FIGURE 1-3. It's the middle of summer, yet a line of shivering visitors clamors to enter the Warming Hut at Crissy Field as the cold sea breeze funnels through the Golden Gate in the background.

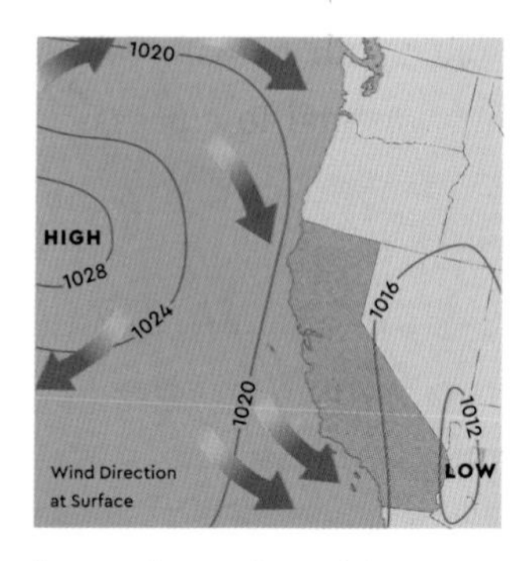

FIGURE 1-4. Summer's prevailing surface winds regularly blow out of the North Pacific Subtropical High, across the cold California Current, and toward the thermal low over our hot deserts.

Air parcels descending out of summer's dominating upper-level high pressure are compressed and heated, settling like a cap above the sea breeze. The resulting warmer-air-above/cooler-air-below conditions may seem odd when you think of how temperatures usually become so frigid at higher altitudes. But the chill of that heavy, shallow sea breeze discourages surface heating down below and holds the marine layer near the surface. These inversions (see figure 1-6) along the coast help define our classic Mediterranean climate; they bolster our summer drought and stamp out any illusions that turbulent storms could loom on the horizon. Many of us climb up California hillsides to suddenly emerge out of the cool haze, fog, and low, flat stratus clouds into the clear, warm atmosphere at higher elevations. We are rewarded with a heavenly view through the dry, warm air with clear summer skies above, while the thick layer of cool, misty atmospheric cotton spreads below.

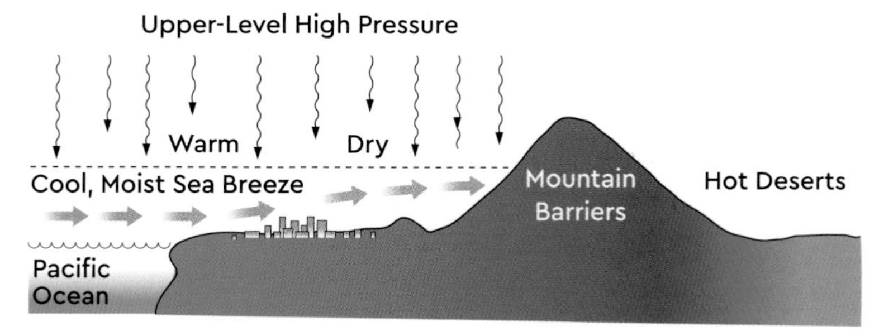

FIGURE 1-5. Particularly during spring and summer, cool marine layers are often trapped below regional inversions along the California coast.

POWER OF THE GYRE

Ocean water temperatures have direct impacts on weather patterns that can spread far inland through the western states and beyond. Our dominating North Pacific Subtropical High pushes out winds that churn the great North Pacific Subtropical Gyre, the colossal clockwise-spinning pinwheel of ocean currents that govern surface water conditions and air masses on the grandest of scales. Draw a line of latitude across every major continent and ocean at just above 30 degrees and you will find the same repeating patterns: magnificent ocean gyres transport warm waters from the equator toward the pole in the western oceans, along the east coasts of major continents. As the ocean water clock continues turning (subtropical gyres swirl clockwise in the Northern Hemisphere and counterclockwise in the Southern Hemisphere), cold water is transported along the west coasts of continents, from the poles toward the equator.

Hence the startling differences in climates when comparing the US East Coast to our West Coast. Contrast San Francisco with Richmond, Virginia, and San Diego with Charleston, South Carolina. Only their latitudes match. Sea breezes on those East Coast cities flow off of their warm Gulf Stream, or may be transported across the continent from the Gulf of Mexico. Summer air masses there are often warm, sticky with humidity, and unstable, leading to summer thunderstorms and heavy downpours. Furthermore, as prevailing winds and weather systems generally drift from west to east at these latitudes, they commonly carry relatively milder air off the Pacific and into the West Coast, but hotter summer and colder winter air masses from off the continent toward and over the East

Coast. In California, cool, stable air masses usually must pass over the cold California Current before coming ashore.

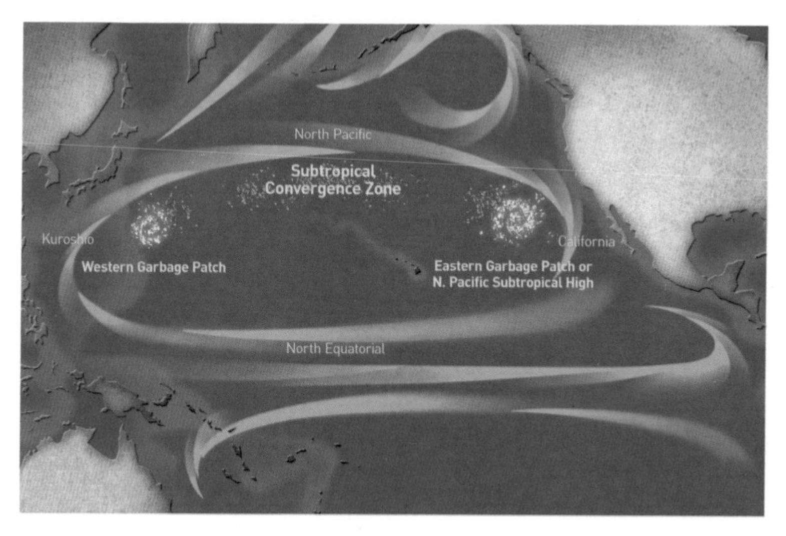

FIGURE 1-6. This simplified map shows the North Pacific Subtropical Gyre that drives cold water down the California coast. It then turns to its right as a gradually warming current while it drifts toward Asia, sending warm water up along Asia's east coast, and will eventually cool again as it drifts north. The pattern is similar in every major ocean basin. Those infamous garbage patches are trapped within the gyre. *Source:* NOAA Office of Response and Restoration.

MARINE LAYER MADNESS

Summer marine layer temperatures near the coast will usually resemble the ocean temperatures where they originated. Fog and low clouds are more common along the north coast because those summer air masses have been in contact with colder water (average water *and* air temperatures are at least 10°F cooler near the Oregon border compared to the Mexican border), keeping air temperatures closer to their dew points for longer periods.

Imagine the puff of fog when you breathe out on a cold winter night: that's your warm, moist breath suddenly cooled to its dew point, or saturation temperature. When coastal (advection) fog streams through the frigid Golden Gate and piles up along north coast slopes, it has condensed in soggy, saturated air blown in from the cold and watery embrace of the California Current. During the day, the low clouds and fog may attempt to drift inland with the sea breeze, above land surfaces warmed by the sun. But the tiny water drops usually evaporate (or "burn off") into warmer air that has gained greater capacity to hold its water vapor. As land surfaces get even hotter, the fog and low clouds appear to "retreat" back over the cool ocean. On some summer afternoons, sea breezes strengthen and the fog races off the cold ocean as if it were attacking the continent, only to quickly

FIGURE 1-7. When you hike up through the coastal sage and above the cool coastal (advection) fog and stratus clouds, you will find yourself above the inversion layer and in the warm, dry air above.

evaporate and disappear over warmer land. Farther south, the air more frequently warms above its dew point, causing the low clouds and fog to clear and making way for those iconic, sunny Southern California summer days. Those same land surfaces will cool off relatively quickly after sunset, allowing the fog and low clouds to spread inland, often covering coastal valleys by late night. The sun will start its warming and "burning-off" process again during the next morning; those repetitive summer weather forecast chants echo along the California coast during late spring and summer months.

Within drier air masses, the atmosphere may not get cold enough to reach saturation. But if the air remains calm on these clear nights, you might observe smaller-scale examples of saturation, condensation, and evaporation cycles when Earth's surface becomes colder than the air. The first surfaces to cool to the air's dew point are usually those that heat up faster and cool off faster, such as metal. Scientists refer to these objects as having low specific heat or heat capacity. A car exterior is a good example. When nighttime air comes in contact with these much colder surfaces, it loses its capacity to hold water vapor and becomes saturated (as it often does over our cold ocean). Condensation occurs, depositing liquid water on those colder surfaces. As surfaces continue cooling overnight, the dew may spread to cover more objects, especially on the cold ground. The same process deposits water from the air onto the outside of the cold glass that holds your icy drink. During California's long seasonal droughts, this early morning moisture can become a lifesaver for many plants and the animals who pause for a lick here and a sip there before the dew evaporates into the warming day.

On a grander scale, these round-the-clock cycles orchestrate similar alterations within moist coastal air masses. During warmer afternoons when air temperatures rise far above dew point, evaporation occurs into air that has a greater capacity to hold water vapor; this air with lower relative humidity feels drier. Overnight temperatures cool closer to the air's dew point, raising relative humidity again toward 100 percent (saturation). And so we see a sort-of daily do-si-do, featuring temperature and relative humidity as the oscillating dancers: when one goes up, the other goes down, and vice versa. Even though the low clouds and fog may retreat each day and return each night during this dance, the marine layer remains in place throughout most summers, often spreading several miles inland in those regions where mountain barriers don't block it. These low marine layers also contribute to milder temperatures, with their blankets of moisture that filter solar radiation during the day and then trap infrared radiation from escaping at night. A comfortable combination of cool afternoon sea breezes and nighttime quilts of insulation conspires to produce California's mildest coastal climates.

Summer's sea breezes and marine layers will also fluctuate within longer cycles that may last for days or weeks. When high pressure strengthens and moves over us, it squashes the fog toward the Earth's surface, limiting it to the immediate coast for days. When high pressure weakens or moves away, marine layers can invade inland valleys and thicken to a few thousand feet. Dense, low clouds may then spread from the coast all the way up against mountain slopes. The changes can be particularly dramatic following inland heat waves, as upper-level high-pressure cookers suddenly weaken or retreat, opening the door to sudden and more powerful cooling sea breezes.

When I was a kid growing up in Santa Ana, I hated the marine layer. I saw it as a dull gray confining slab, obstructing the expansive sky that I wanted to view. And when it finally "cleared" to haze, I often couldn't see very far through the smog. Perhaps it reminded me of my confinement to a place where I didn't want to be. By contrast, clear skies represented a window to the world I wanted to explore. I wouldn't be the first kid with such an imagination and craving for adventure. That's probably why I learned to love warm or unsettled weather. No marine layer, and you can see all the way to the mountains and any significant clouds or weather that might be popping up near the horizon.

Today, I appreciate the marine layer. I'll take 75° over 99°, especially when I have to live, work, and play in such weather. I remarked to a friend the other day that our air near the coast was so soft, it was like being surrounded by fluffy pillows. Millions of others have noticed, too, and residing in that soft, fluffy pillow air along our coasts has become exorbitantly expensive.

Nippy summer breezes and thin low clouds and fog along the coast may create an illusion that the sun's rays are harmless, or not even making it to the surface. But almost all of the incoming ultraviolet radiation sneaks through these thin veils of chilly maritime moisture, burning the shoulders and noses of forgetful locals and those unsuspecting visitors more familiar with the immediate and intense heat along the Atlantic shore. You might blame ocean currents for these differences between East Coast and West Coast summers.

The cold California Current differs dramatically from the south-to-north-flowing Gulf Stream, which carries relatively warm ocean water up along the East Coast. The Gulf of Mexico and Gulf Stream water temperatures can soar above 80°F (27°C) by late summer. Warm, muggy, and therefore unstable air forms over the warm currents off the southeastern US. By contrast, late-summer water temperatures may finally peak at only about 70°F (21°C) along the Southern California Bight; they often remain in the 50s north of Pt. Conception, even into the warm season. (Pt. Conception teams with Pt. Arguello to form that massive elbow protruding into the Pacific, effectively separating Northern and Southern California coastal waters. The Southern California Bight refers to the more protected coastline trending southeast from there to the Mexican border.) Such cool ocean currents keep our summers not only cool but relatively calm, compared to a state such as Florida, which is the thunderstorm capital of the US.

FIGURE 1-8. Temperatures remain near 70°F (21°C) in the air and water on this beach near San Onofre, even though the sun is high on this summer day. Typical low stratus clouds are "burning off."

BLOCKING THE SEA BREEZE

Air masses flowing off our cold ocean are modified as they move inland. Traveling from the cool beaches to overheated inland valleys, you will experience some of the most extreme summer afternoon temperature gradients in the world, occasionally exceeding 1°F per mile.

Topographic barriers represent major sea breeze blockades. This is why parts of Long Beach downwind of the Palos Verdes Peninsula's hills may run 10° warmer on summer afternoons, compared to nearby beach cities. When strong high pressure builds overhead and squashes the marine layer closer to the surface, world-famous Malibu and Topanga beaches often experience 70°F (21°C) summer afternoons, while less than 10 miles inland and over the mountains, Woodland Hills and Calabasas are roasting in temperatures well over 100°F (38°C). Pick your favorite adjacent coastline, mountain range, and inland valley from Mexico to Oregon and you may discover similar temperature contrasts, especially throughout the summer.

Around the Bay Area, you will find startling temperature gradient examples across Marin and San Mateo Counties. Traveling from the beaches on the western sides of their mountains to the inland (leeward) sides during summer afternoons, you will often transition from wintery dull gray to summer's sunny warmth. The world's bustling epicenter of technology—San Jose and the Silicon Valley—was built in a basin, the Santa Clara Valley, protected by the Santa Cruz Mountains from direct hits of maritime air masses. This leaves the valley with consistently warmer and sunnier summer afternoons compared to the cold coastlines from San Francisco to Monterey Bay. Travel farther south on the inland side, more distant from the cooler

San Francisco Bay air masses that may have swirled through the Golden Gate and down into the Santa Clara Valley. Where orchards once relied on the regularity of sunny summer weather, you will encounter increasingly higher summer afternoon temperatures through the sprawling suburbs that have replaced farms all the way south past Morgan Hill and Gilroy.

Majestic, rugged ranges rising along the far north coast are also formidable barriers to shallow sea breezes and marine layers. Summer heat can rule surprisingly close to the coast, within narrow valleys and canyons that become sheltered from sea breezes as they meander inland through steep mountain barriers. For the foggiest summer, find a cabin on Cape Mendocino or the Lost Coast, both of which stick out into persistent, frigid fog banks. Patches of coast redwood forest (or places where they grew before they were logged) are usually found in areas with the most consistent summer fog. "Where the fog goes, the redwood grows"—the cool coastal mist not only decreases evapotranspiration (the total amount of water evaporating from plants and other surfaces) in these plant communities but also adds more than 10 inches of fog drip to the forest floor during summer's drought months. Heavy rains take over during winter, nourishing the natural range of coast redwoods and other mixed coastal forests from Big Sur north past the Oregon border. But you won't find redwoods growing in natural stands far inland and distant from the fog belt.

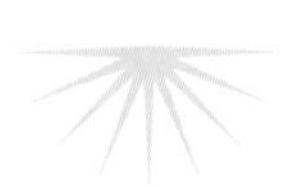

BASEBALL WEATHER . . . OR NOT

When the sweet smell of freshly cut grass floats through the air on a long sunny day or beneath light towers in the summer night's breeze, the season for baseball is finally here. Rain, wind, and cold are softball's and baseball's dreaded enemies. Why do so many Major League Baseball players come from California and its baseball-championship schools, especially in Southern California? Because fair weather days leave plenty of time to hone their baseball skills. There are still some winter and early spring days when the turf becomes saturated and squishy (particularly in Northern California), or when the oppressive tule fog limits visibility in the Central Valley, or when a bitter blustery wind whips across the fields of dreams to kick up dust in the high desert. Every batter dreads the excruciating stinging sensation that reverberates into their hands and up into their arms when the ball hits an awkward spot on the bat during cold spells. But those inclement breaks are relatively brief compared to the longer mild stretches of baseball weather that dominate from spring through autumn, fair weather that is more likely as we move south in the Golden State. Baseball weather has helped to nurture five major league teams in the state. I've enjoyed watching baseball and the weather in all of their stadiums. Our best big baseball parks (and smaller baseball and softball diamonds) are positioned to open out with the prevailing wind so that fly balls are launched with the wind (and often the sun's rays) at our backs instead of into it. Changing sun angles during different times of day and year are also important factors when positioning ballparks.

The Los Angeles Dodgers' baseball stadium is perfectly situated on the hill above downtown to take advantage of summer

FIGURE 1-9. Cool temperatures, high relative humidity, and fog drip enable Northern California's coast redwood (*Sequoia sempervirens*) to survive long summer droughts.

weather patterns. There, fans can view games framed by swaying palm trees in the foreground and the scenic San Gabriel Mountains in the background. The stadium is located far enough from the beach to warm in the summer sun, but not so far from the ocean that it roasts in the consistent oppressive heat of the inland valleys. Likewise, Angel Stadium of Anaheim is located along the Santa Ana River in inland Orange County, between the cooler coast and the hotter inland valleys, also looking toward the mountains, perfect for baseball weather. The San Diego Padres' baseball park (facing more toward the north) is conveniently plopped right in the middle of downtown, where summer's refreshing sea breezes flow directly off of adjacent San Diego Bay. Northern California presents a few more baseball weather challenges. The old Oakland Athletics' stadium (since 1968) always lacked a baseball feel. But it was located on the east side of the Bay so that it didn't suffer consistent direct hits of bitter winds off the cold Pacific, and it boasted reliable summer temperatures rated between cool and mild. Nevertheless, this circular park by the Bay ranked as one of Major League Baseball's breeziest. As of this writing, the team is settling on final proposals for a new stadium in Las Vegas. The Oakland A's could say goodbye to generations of fans as well as to mild summer sea breezes.

The San Francisco Giants Major League Baseball team has established memorable and intimate relationships with its sea breezes and fog banks. Placing a baseball stadium in the path of San Francisco's storied summer fog and biting winds is a miserable mistake for players, spectators, and a baseball franchise interested in profits. This is one reason why today's stadium is not located at Crissy Field, in Golden Gate Park, or in the Sunset or Richmond Districts, where frigid summer winds whip avenues of summer fog directly off the cold Pacific. But it also makes you wonder why the team's old stadium, Candlestick Park, built in 1960, was placed in windswept Bayview Heights near Hunter's Point. Four decades of baseball circus acts followed because someone did a poor job of analyzing local weather patterns.

Strong west–east winds that blow off the Pacific and across Lake Merced funnel into Alemany Gap, where they must split around Bayview Hill and its park, only to swirl around and meet again on the park's lee side next to the Bay. Who decided to assemble Candlestick Park in this wind-rendezvous southeast corner of the City? Many night games became trials of survival for all. Fans huddled under warm blankets and parkas all year long. Baseball's finest defensive stars were humiliated by fly balls launched into the swirls of shifting erratic gusts; no person could accurately predict a final outcome until the mischievous baseball finally made its surprise landing in a glove or on the ground. The difference between official hits and errors became nebulous and controversial, sparking debates and jokes as winds whipped these games into entertaining, hilarious, and thrilling victories or maddening agonies of defeat. No wonder Tony Bennett sang about leaving his heart in San Francisco above the blue and windy sea.

The Giants, in 2000, finally and mercifully moved their games to their now-beloved stadium closer to downtown. It is nestled in China Basin, protected on the sunny east side of town by nearby hill barriers, where batters can drive home runs into the breezy Bay on brilliantly sunny summer days and cool summer nights. Although situated in one of the warmest parts of the City, the stadium still ranks among the most consistently cool and breezy Major League Baseball parks in the nation during summer. Jubilant sold-out crowds have voiced their approval. All along the California coast are similar, though less storied, local wind and fog gaps, funnels, and barriers that break through the coastal wall. Smaller baseball stadiums are scattered around less populated parts of California, too, hosting minor league teams with loyal local fans who put up with oppressive summer heat in inland valleys instead of freezing in the Giants' bleachers.

FIGURE 1-10. Summer fog rushes inland from the Avenues and the Golden Gate, but then must detour around the hills that helped make San Francisco famous. It evaporates in the warmer air as it drifts inland with the sea breeze.

THE MARINE LAYER OWNS US

I laughed when I first heard some local National Weather Service meteorologists coin the phrase, "The marine layer owns us." They were jesting about how very subtle local variations in the height and intensity of invading maritime air can determine the difference between a cool, gray day and a warm, bright, sunny afternoon along the coastal plains. Dramatic temperature differences over short distances are not limited to summer months. This effect of mild marine air versus extreme continental air may also rule during winter nights and early mornings, sometimes creating similarly dramatic temperature gradients, but in reverse. Just as they heat up faster, land surfaces cool off much faster than relatively invariable water during long winter nights, leaving coastlines milder than slopes and canyons that face inland and lack such strong marine air influence.

FIGURE 1-11. Strong high pressure scours out the marine layer along the Southern California coast on August 5, 2021, while fog and low stratus clouds hug most of the central and north coasts. Southerly winds rotating clockwise around the inland high push smoke from historic wildfires into the Pacific Northwest. *Source:* NOAA/National Weather Service.

Coastal and inland Californians experience such wide variations in temperatures due to the differences between water and land. It takes a lot more solar energy to increase the temperature of the same mass of water than it does land. Water must absorb about five times more energy to heat up as fast as soil. (Scientists refer to this phenomenon as high specific heat capacity, which allows water to store and release tremendous amounts of energy even with the slightest changes in temperature.) Also, solar radiation is absorbed at only the very thin surface layer of the land, but it penetrates and is distributed deep into water. Further, water currents circulate and distribute heat energy throughout water bodies, whereas thin, motionless layers of land surfaces remain directly exposed to sunlight and air temperature changes. Finally, when water masses heat up, the warmest water evaporates, taking its latent heat with it and leaving behind cooler surfaces. (Latent heat is absorbed as liquid water evaporates; this energy is stored in water vapor, but can't be measured by a thermometer.) These factors combine to cause land surfaces to heat up and cool off rapidly, while mild ocean temperatures (and the air masses above them) exhibit relatively small thermal variations. Conspiring with these factors, dry air masses over the land allow direct sunlight to blast to the surface, but that same dry air allows infrared heat to quickly escape at night. By contrast, moist air masses off the ocean act as shields, protecting surfaces from more intense shorter-wave solar radiation by day. At night, these same wet blankets absorb longer-wave infrared radiation from Earth's surface and then radiate it back to us. You know the marine layer mantra by now: shields during the day, insulating blankets at night.

Air masses forming over and flowing off of land surfaces—offshore breezes—tend to be hotter during the days and summers, but colder during nights and winters. By contrast, sea (onshore) breezes tend to cool us off in the summer and remain relatively mild during winter. Earth's mildest climates are found on islands and along coastal regions with prevailing sea breezes, such as coastal California. Our West Coast prevailing sea breezes typically transport milder, moister air masses into our state from the west.

I have been fortunate to have experienced each season throughout very different regions of the state, ranging from its foggy north coast to its higher-elevation remote mountain extremes to its searing deserts. I was always attracted to these places with more extreme weather because they were most dissimilar to where I grew up, which was only 10 miles from the beach (though it seemed much farther). These combined life experiences in other places have further helped me grow to better appreciate the comforts and conveniences of living along the Southern California coast. What I once scorned as boring weather monotony has evolved into celebrating the daily advantages of living in one of the world's mildest climates, where you can plan just about any outdoor activity without worrying about interfering weather extremes. Revisiting coastal California climates during summer reminds me how we can become accustomed to the ordinariness of our daily lives, and how by leaving and then returning we can restore a sense of the wonder that was always there, hidden under the layers of familiarity.

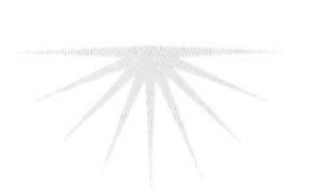

PUSHING INTO HOTTER INLAND VALLEYS

Summer's stable sea breezes carry us now deep into the Golden State's inland valleys, to the bases of our major mountain ranges. Plants and animals (including humans) must adapt to the harsh temperature extremes typical of these inland locations. By the time it reaches regions such as the Central Valley and Southern California's Inland Empire, the already diminished marine layer meets even larger obstructions compared to the lesser hills and mountains it traversed to get there. These giant mountain ranges (such as the Sierra Nevada and the Transverse and Peninsular Ranges) act as atmospheric barriers as they rise up more than 10,000 feet (3,000 m) to block the thickest maritime air masses. Smog transported from millions of human sources is also trapped in these stable layers of air, below summer's high-pressure dome. When the pollution is baked in the sun, a series of chemical reactions produces toxic ozone. Measurements show dramatic peaks in this photochemical smog during the afternoons, just when many people are most active. An oven-like summer calm and musty, dusty heat often settle in the San Joaquin Valley, accumulating smells of crops, cattle, and the exhaust from the many huge machines that help grow and transport our food. In every inland valley, temperatures drop and the ozone dissipates later at night.

During many summer afternoons, as valley surface air is superheated, the air expands and lifts the inversion and the air pollution up warmed mountain slopes and into the canyons. High concentrations of ozone and other pollutants are carried toward foothill and mountain resorts, national forests, and national parks. This air pollution has weakened and killed millions of trees and damaged entire forests along the western slopes of the Sierra Nevada and on mountain slopes ringing the LA Basin. Sequoia seedlings are particularly vulnerable. Partly due to fewer pollution sources and greater atmospheric mixing, the Sacramento Valley to the north usually records better air quality than the San Joaquin Valley. An unfortunate exception to this in recent years comes in the form of smoke from more frequent wildfires in the surrounding mountains. In an unforgiving cycle, our air pollution weakens and kills trees, creating tinder-dry fuel that burns to produce more air pollution.

There is also some good news. From the 1960s and into this century, California's population and the number of cars doubled, and its economy grew by many more times. During that period of a few decades, however, smog levels were cut in half in much of the Golden State (including LA), proving that healthier air can coexist with a growing economy, and dismissing the archaic falsehoods that pit jobs against the environment. Concerted efforts to become more efficient and to clean up pollution have saved billions of dollars that would have otherwise been wasted on dirty, inefficient energy sources and the staggering health care costs of treating people who have been sickened by toxic air. Millions of people have also been saved from suffering from debilitating respiratory illnesses. There remain plenty of opportunities to improve California's air quality, but there are countless healthy kids and scores of others who are still alive, thanks to efforts to understand how we can prosper from healthy air *and* a healthy economy.

In contrast to periods when the inversion is very shallow, weaker or retreating upper-level high pressure will allow summer's sea breezes to expand and strengthen. Dilution is one solution to pollution, and smog is better dispersed as it mixes higher into taller air columns during these cooler periods to give us all more breathing room. This air will become warmer and drier as it finally squeezes through Southern California's Tejon, Soledad, Cajon, and San Gorgonio (Banning) Passes and into the deserts.

To the north, cool relief may creep deeper into the Salinas and Santa Clara Valleys, warming along the way. East Bay cities nestled in inland valleys behind the hills will also enjoy cooler summer days with less air pollution when the classic inversion lifts to higher altitudes, allowing assertive sea breezes to spread east all the way into places such as Livermore. The strongest sea breezes channel through the Golden Gate and Carquinez Strait and then pour out across the Sacramento–San Joaquin Delta. This Delta breeze can bring welcome and refreshing breaks to summer's heat waves far into the Central Valley.

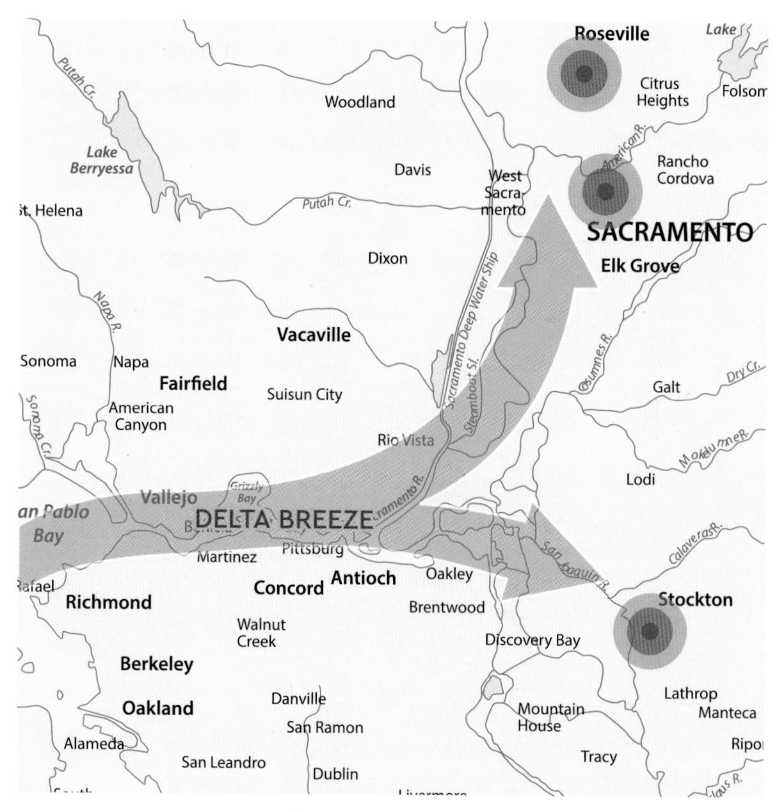

FIGURE 1-12. Summer sea breezes blow cool air inland, splitting north into the Sacramento Valley and south into the San Joaquin Valley.

On most summer days, this Bay-to-Delta breeze and its remnants have already been heated and dried out as the breeze splits to the north into the Sacramento Valley and to the south into the San Joaquin Valley, thoroughly blocked by the giant wall to the east that we call the Sierra Nevada. The river-of-air branch that turns north, over Sacramento and toward Redding, can bring penetrating relief from summer's otherwise searing heat. But the heat dominates as we move farther north away from the Delta until temperatures in Redding (boxed in by mountains at the far northern tip of the Sacramento Valley) climb well above 100°F (38°C) on numerous summer days. (Redding's July average high is over 98°F.) This north end of the valley is only 100 miles east of the cold ocean, where summer afternoon temperatures rarely make it to 70°F (21°C) on the beaches. But

FIGURE 1-13. Giant sunflowers and other crops flourish in the hot summer sun of the Sacramento Valley, but most require at least some irrigation.

the mighty Klamaths and North Coast Ranges block onshore breezes, keeping those foggy ocean air masses from penetrating too far into coastal mountain canyons. By the time the air passes around topographic obstructions and mixes with inland heat, it may have already stalled and warmed above 100°F (38°C) within local valleys, long before approaching the massive Sacramento Valley. Residents in this valley surrounded by mountains on three sides must look toward the distant Delta for relief from summer's fever.

Likewise, the summer river-of-air branch that veers south out of the Delta toward Fresno and then Bakersfield might moderate temperatures until it loses its momentum and mixes into the sizzling San Joaquin Valley's pool of summer air. Here is where we recognize one of those magical coincidences in

nature. Consider how streams and rivers flow downhill and out of the Sierra Nevada and Northern California, converging into the Central Valley, all flowing roughly parallel to summer's afternoon surface winds, but in exactly opposite directions. This is melt-water flowing from winter's dwindling high-country snowpack and from stored water released out of the many reservoirs that distribute this precious liquid gold to farmers and other users throughout summer's drought season. Just as beachgoers are relaxing and reveling on strips of sand along the cool coast where the sea breezes originate, so are Central Valley residents finding open strips of sand along their rivers and streams that are flowing toward the coast. While the coastal cohort plunges into powerful ocean waves and shivers in a cool sea breeze flowing inland, valley residents might dip into refreshing river and stream water currents that can soothe the discomfort of summer's inland oven . . . all while their river water flows toward those colder beaches. Eventually, these waterways converge at the Delta, where runoff is squeezed through the Carquinez Strait and then spread into the bays until it is funneled through the Golden Gate and into the ocean, as if to mimic the sea-breeze rivers of air, but in reverse.

FIGURE 1-14. On many summer days, surface heating conspires with on-shore flow to push an otherwise shallow inversion (and trapped smog) from inland valleys up and into mountain canyons (left to right here).

HARNESSING THE WIND

As you follow the prevailing sea breeze, heading inland, you'll often reach a pass where wind turbines tower hundreds of feet overhead like enormous industrial sunflowers. Their clusters mark the transition from the milder coastal climates into the harsher temperature extremes farther inland. Erected where the most impressive funnels link the coast to summer's roasting valleys and deserts, they catch marine air racing inland to fill the void of thermal low pressure that forms in the hot, expanding air on the continent. (These passes may also funnel fall and winter offshore winds in the opposite direction.) With these prime locations and conditions, California leaped ahead in harnessing wind energy, and is the first state where large wind-power stations were developed. The entire state's average seasonal wind-power production increases toward annual peaks in the spring and into summer, thanks to such consistent breezes that are channeled through the mountain passes. As technologies improve each year, California's varieties of towering wind turbines combine as national leaders in wind production, and wind continues to be one of the most economically efficient and fastest-growing sources of electricity in the Golden State.

Most notable are the Shiloh Plant in Solano County and the larger Altamont Pass, where breezes flow from the Bay and spread into the Central Valley. Tehachapi Pass above Mojave and San Gorgonio (Banning) Pass (spotlighted in a section in chapter 4) are the leaders in Southern California. Smaller power stations have sprouted throughout the state, anywhere there's a prevailing wind and a mountain pass to catch it.

Where winds are most reliable, the turbines are transforming landscapes; they are even impacting and threatening species of migrating birds, which sometimes ride on the same wind tunnel currents where the turbine blades are rotating. At stake are migratory patterns that have been established over millions of years, as birds noticed the effect of mountain topography on wind direction and speed and used it to quicken their journeys and lessen their own energy expenditures. Numerous research projects championed by universities, government agencies, wildlife organizations, and energy companies continue to add to the volumes of what we have already learned about this problem. For example, organizations such as the Center for Biological Diversity, BirdLife International, and Audubon have summarized how thousands of birds and bats have been killed by Altamont Pass turbines every year, including hawks, kestrels, and golden eagles. Surviving birds are forced to search out safer perches and routes. Recent improved placement strategies and turbine designs will cut the losses.

FIGURE 1-15. These towering turbines are part of the Shiloh Wind Project in Solano County, about 40 miles northeast of San Francisco. Here, they rise above ranchlands in the Montezuma Hills, where winds from the coast sweep into the Central Valley.

WORLDS AWAY FROM THE OCEAN

Our summer adventures now take us farther from the coast, into our most extreme inland climates that often seem planets away from the state's iconic beaches and coastal valleys. At lower elevations, we find blistering summer heat in the continental air masses on landlocked sides of the major mountain barriers. At more lofty elevations, we are high above summer's marine layers and desert heat; winter's icy grip has given way to the spectacularly warm days and comfortably cool nights that attract throngs of visitors, migrating birds, and other animals to our mountain resorts. These tallest ranges and ridges trend mostly north–south or northwest–southeast, from Oregon across the Mexican border. These monumental walls separate what is sometimes called cismontane (landscapes facing toward the coast) and transmontane (landscapes facing toward the continent) California. You can roughly follow the top of this enormous air mass blockade by hiking the Pacific Crest Trail. From a High Sierra perch looking west toward the coastal horizon, the weather-aware backpacker will observe maritime air masses; turning to look inland and east means facing drier and harsher continental air masses exhibiting greater seasonal and diurnal temperature extremes. Examples of radical temperature extremes common to transmontane California abound from north to south, almost always on the continental side of the great barriers. To best observe the brightness of the Summer Triangle—comprising the stars Altair, Deneb, and Vega—and summer constellations among a tangle of other stars in pristine dark night skies, go to the high and dry places, where highly variable climates contrast with the hazy constancy and urban light pollution of the coast.

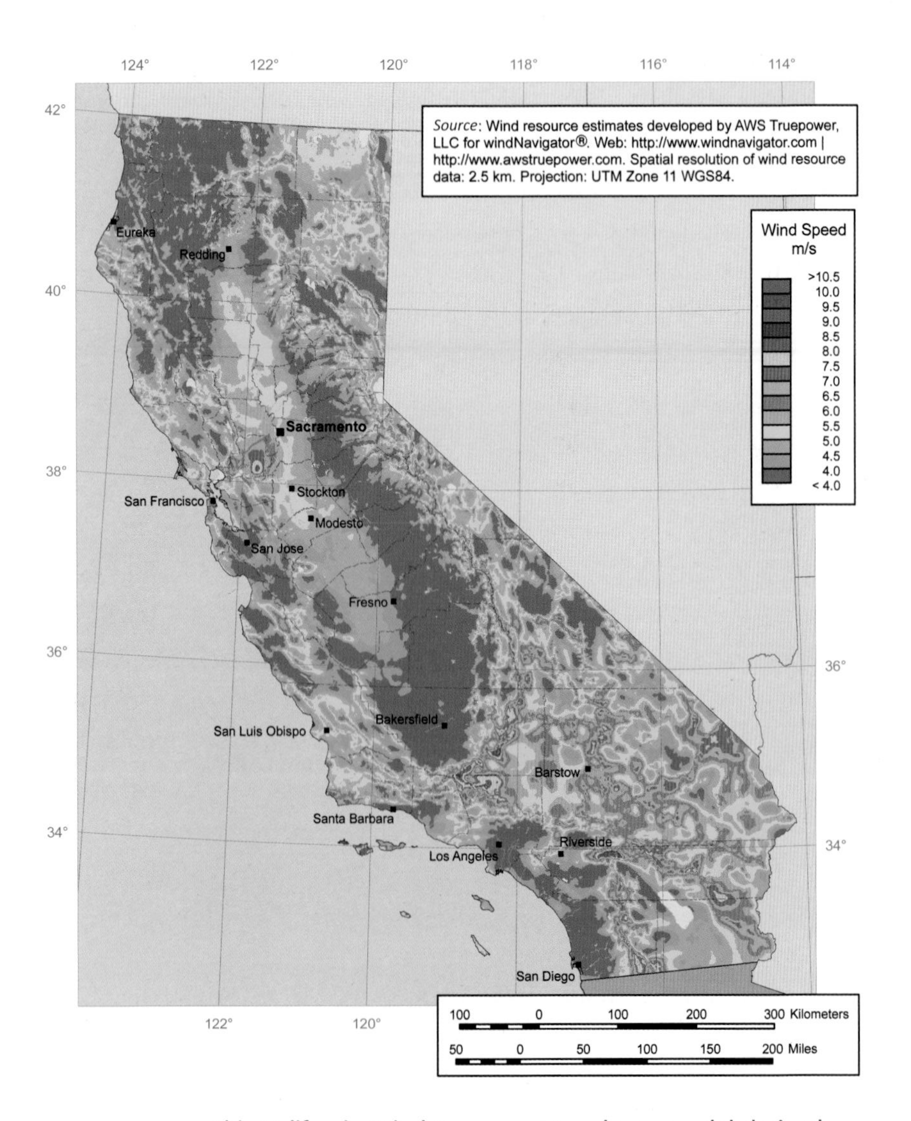

FIGURE 1-16. This California Wind Resource Map shows model-derived estimates of average wind speeds at about 262 feet (80 meters) above the surface, the height of many of today's large wind turbines. The threshold for suitable wind development is around 6.5 meters per second (about 14 miles per hour). Note how many prime locations are concentrated in local passes. *Source:* U S Department of Energy WINDExchange: https://windexchange.energy.gov/maps-data/12.

Low elevations of the Coachella, Imperial, and Colorado River Valleys of Southern California become massive summer ovens, for three primary reasons: intense solar radiation rapidly heats dry desert surfaces during long days with high sun angles; air masses descending from tall stacks of persistent dominating high pressure aloft are heated by compression, trapping the fever near the surface; and mountains to the west block any intrusions of cool ocean air, while winds cascading downhill from those same mountain slopes may be further compressed and heated. Here near the Mexican border during the summer solstice is where the noon sun angle peaks near 81 degrees from the southern horizon, just 9 degrees from the top of the sky; but there are no cooling sea breezes so common to the coastal side of the mountains. Temperatures can soar to more than 120°F (49°C) in these valleys that are often considered extensions of the more expansive Colorado, Sonoran, or low deserts. When summer afternoon temperatures may struggle to reach 70°F (21°C) in Oceanside, they may soar well past 110°F (43°C) in Palm Springs, less than 100 miles away, but on the inland side of the towering San Jacinto Mountains. Temperature differences between the beaches of Orange and San Diego Counties and the Imperial and Coachella Valleys may reach 50°F (28°C) on the most extreme summer days.

But you will find the hottest temperatures in that distant deepest valley, far across the Mojave Desert and into the Basin and Range Physiographic Province. Death Valley has earned its name and reputation: at 282 feet (86 m) below sea level, it is the deepest basin in North America. It is Earth's perfect roasting pan. It bakes near or below summer's relentless upper-level high-pressure ridges; as the crow flies, Death Valley is less than 300 miles from the coast, but is four major mountain barriers away. Maritime air masses struggle to penetrate those most influential blockades that divide California: the Coast Ranges and the Sierra Nevada. Even if ocean winds could get that far inland, there remain two more giant walls to navigate. This blistering lowest valley is also surrounded by its own impressive mountain ranges, requiring all invading air parcels to descend to its depths, blocking any remnant moisture and forcing air masses to heat by compression before they can reach the bottom.

Thus the official Death Valley weather station repeatedly records the hottest temperatures in the nation and sometimes on Earth during long summer days. From late June into August, afternoon highs can soar well over 120°F (49°C), and overnight lows may not drop below 100°F (38°C) during some of summer's warmest nights and mornings. Here, the hottest temperature recorded in North America—134°F (56.7°C)—has recently been scrutinized as the highest officially recorded temperature on Earth. This hotly debated record has been threatened in recent summers, when the valley's official thermometer pushed toward and even touched 130°F (54.4°C) in 2020 and 2021. Perhaps the record has been breached by the time you read this. Nature's infamous annual bake-out has even attracted summer tourists who want to check it off their list of extreme weather experiences. Busloads of visitors from Europe, Asia, and other distant lands, on their way to Las Vegas or LA or San Francisco, have taken a detour into this remote valley during the scorching season. They are pioneering "heat tourism" just to sense summer's continental tropical air masses.

Every visitor to these dangerous desert ovens is warned ahead of time of what local animals have already learned. There is far more at stake than oppressive discomfort. A person venturing too far from their bus or vehicle or shelter may never return. Dehydrated bodies of careless hikers are discovered too late after they wandered out to worship the heat that quickly killed them. The combination of dry heat and intense sunlight can draw and then evaporate gallons of water out of the body within an alarmingly short time.

When introducing students to the desert, I planned all my field classes so that my students were doing their work during fall, winter, or spring of each year and not risking the travails of midday summer. I have purposely tested the oppressive heat in this frying pan of a basin, but I don't recommend it to the fainthearted. For my own education, I have walked out into its sinuous mountains of sand dunes that I had previously enjoyed during cooler times of year. On one July morning, moonshine was gradually consumed by dawn's early light. This is the time of day to look for the hardy beetles, rodents, rabbits, snakes, road runners, coyotes, and other animals that make their home in the desert. I was careful to keep my bearings and flee the sandy dune roasting pan after the day's torch appeared and began rising on the northeastern horizon. The overnight/early morning low during this visit was only 98°F (37°C). By the time I and the very few other human visitors had fled to the "safety" of our vehicles, the already blistering dunes seemed abandoned; the animals had escaped into their natural shelters, where they would huddle until long after the day's sky fire finally set in the west. Experiencing summer's discomfort in California's deserts helped me better appreciate and respect the common creosote and other desert scrub species that manage to survive and endure, even when exposed to a punishing sun that seems certain to dehydrate every bit of life out of every organism.

FIGURE 1-17. The desert tortoise lives underground during most of its long life, sometimes emerging during spring or after soaking rains. It also survives the harsh climate by storing fat and water.

Several California desert animals carry unique adaptations to help them survive temperature extremes and water scarcity. They offer lessons about adjusting to severe weather and demonstrate the resiliency of life under adverse conditions that would kill most organisms. Many of these acclimations involve avoiding the extremes without relying too much on water-consuming evaporative cooling. The Mojave desert tortoise (*Gopherus agassizii*) is just one example among the more than fifty reptile species found in California's deserts. This herbivore can live up to eighty years. It avoids extreme weather by remaining dormant underground for up to 95 percent of its life, emerging during spring or after a soaking rain and by carrying reserves of stored fat and up to a quart of water, mostly concentrated in its urinary bladder. These laid-back tortoises have survived in and adapted to some of these most extreme climates for thousands of years. Unfortunately, increased harassment from humans and predation by ravens and many other introduced species have made the formerly common but now threatened desert tortoise a rare sight.

Among the rodents are the different species of kangaroo rats (*Dipodomys* spp.). They may live their lives without ever drinking water: they conserve moisture derived from seeds, and their hyper-concentrated urine is almost crystalline. Temperatures remain moderate and humidity relatively high within their sealed burrows. You might also find black-tailed jackrabbits (*Lepus californicus*) sheltering in the shady north sides of shrubs on a hot day. Their lanky limbs and giant long ears are efficient radiators of excess heat. Look for a coyote (*Canis latrans*) repeatedly pouncing and flushing a rabbit from shrub to shrub until the prey overheats enough to offer an easier meal. The panting coyote uses evaporative cooling from its tongue, just as we rely on evaporative cooling from perspiration on our skin, but this process releases a lot of precious water.

FIGURE 1-18. Death Valley's dunes should be evacuated as soon as the summer sun appears. Note the large alluvial fan in the background. It was formed as successive mud and debris flows were coughed out of the mountain canyon just above the fan during rare, intense summer thunderstorms.

FIGURE 1-19. Cholla and other plants and animals have adapted to the harsh temperatures and extreme drought common to the Mojave Desert. This photo shows a few summer cumulus clouds that have boiled up within afternoon thermals, casting their merciful shadows across the nearby New York Mountains.

Among the numerous bird species, roadrunners (*Geococcyx californianus*) may be best adapted to desert extremes. They have developed a variety of behaviors to save energy and to moderate body temperature, which include sunning themselves on cold days. They look for the densest nurturing foliage to make their nests, where the number of eggs laid will depend on the abundance of food and water during that season. They are among the most efficient water-saving birds. They may even eat weaker nestlings that have little chance of survival, demonstrating that drastic measures may be required to conserve resources and propagate a species in these punishing climates. I have watched Death Valley roadrunners sheltering under the shade of creosote, then briefly emerging into the parking lot to pick dead bugs off the front bumpers of vehicles, as if humans had been collecting for their buffet.

MOVING ON UP

Because Death Valley is the lowest place in North America and is surrounded by mountains that block potentially cooler and wetter air masses, it represents the baseline for California's most extreme summers. So let's travel up from this desert floor to see how quickly the weather can change. Temperatures decrease by an average of about 3.5°F per 1,000 feet (6.5°C/1,000 m) as we move uphill, ascending the badlands, alluvial fans, bajadas, canyons, and mountain slopes that define these spectacular desert landscapes. By the time we reach the heights of Telescope Peak, just above 11,000 feet (3,350 m), summer afternoon temperatures are often 40°F degrees cooler than the salty, below-sea-level desert skillet basin shimmering beneath us but still within our view. Near the peak, protected patches of snow may linger toward summer when temperatures are soaring over 120°F (49°C) at the bottom of Death Valley. We are now weather worlds away from the mild maritime coastal climates where we started this journey. But this is only one location. The tremendous range of elevations across our tall-mountains-next-to-deep-valleys state creates an enormous diversity of weather patterns and climates.

In the relatively cooler and wetter high deserts, and then further up into the Golden State's revered mountain environments, we encounter very different atmospheres and ecosystems. California sometimes records the coldest overnight low temperatures in the nation (excluding Alaska), thanks to these high and dry locales. The weather station in the high-desert ghost town of Bodie is a frequent winner. Wildflower season, which starts in late winter at lower elevations along the coast and in the lowest desert basins, migrates north and into higher mountain terrain as spring progresses toward summer. By late summer, heat and drought have encouraged the wilting blooms to advance farther north and up to the loftiest, coolest mountain slopes, just after the seasonal snow blankets have melted and just before the next season's frosts and snowstorms return. When you follow this annual wave of seasonal wildflower displays, it will guide you north or higher up into our sky islands with their cooler climate zones . . . until you run out of terra.

Climate, including temperature and the availability of moisture, is often the most important factor shaping ecosystems as they evolve and thrive throughout California. The state's large range of elevations and climates encourages unparalleled plant diversity. Using modifications to C. Hart Merriam's 1890s research in the southwestern US to survey life zones from lower to higher elevations in California, we notice as much variation as if we were traveling from hotter northern Mexico to colder northern Canada. We start near sea level in the hottest, driest Lower Sonoran grasslands and desert scrub. A few thousand feet higher, the Upper Sonoran chaparral, woodlands, sparse pine, or high desert sagebrush may dominate. By about 5,000 feet above sea level (1,500 m), diverse yellow pine forests with species such as ponderosa and Jeffrey pine appear in the cooler, wetter Transition Zone. In the Canadian Zone up around 7,000–9,000 feet (2,130–2,740 m), even cooler, wetter, often denser forests of lodgepole pine and red fir are common. As we reach the tree line around 10,000 feet (3,000 m), subalpine species of the Hudsonian Zone dominate. Above 11,000 feet (3,350 m), we find only the hardiest prostrate species that can survive the harsh, icy winters and short summers of the Arctic-Alpine Zone. Because available water plays a major role in marking these life zone/vegetation zone boundaries, the zones are often found at gradually lower elevations as we move into the rainier climes of Northern California and at higher elevations farther south. Likewise, the zones are usually found at lower elevations on wetter western cismontane slopes and ranges and at higher elevations on drier continental transmontane and rain shadow terrains. On a much smaller local scale, these vegetation zones might even be found lower on shadier north-facing slopes and at higher elevations on drier south slopes facing toward the sun.

By learning to notice where different plants and animals live and grow, you can teach yourself about the weather patterns and microclimates at any particular location. Sizzling low-elevation summer temperatures gradually moderate on the way north and to higher elevations; in far northern California, forests that grow near sea level would only survive at high elevations in Southern California. In her classic book, *An Island Called California*, Elna Bakker traversed the Golden State's amalgam of plant

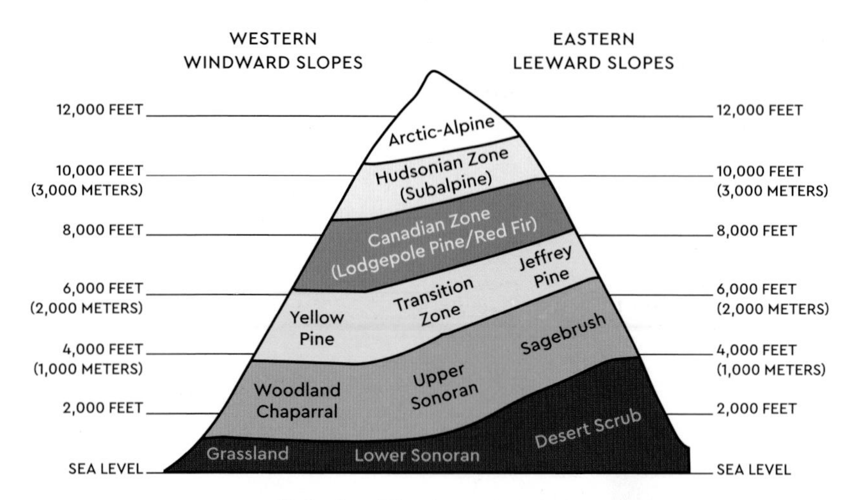

FIGURE 1-20. You will find California's myriad plant communities within these generalized vegetation or life zones. Climate (particularly temperature and moisture availability) is usually the principal factor that determines their locations and characteristics.

communities, reminding us that microclimates are often the most powerful regulators of plant life and resultant ecosystems.

Exhibiting temperature extremes within dry microclimates, the cooler, higher deserts are often decorated with Joshua tree woodlands of the Mojave and then high desert scrub, pinyon, and juniper woodlands that range from the state's eastern mountains well into Nevada. Travel Highway 395 from the Owens Valley up into the Modoc Plateau of northeastern California to find such communities. Summer temperatures over 100°F (38°C) are common; although cooler than in the lower, southern deserts, summers are usually hotter in these high continental deserts than in those maritime valleys far to the west, on the moister coastal sides of nature's major rock walls. As we ascend above 5,000 feet (1,520 m), we emerge into California's still cooler and wetter mountain forests, our sky islands. Even when located closer to the coast, they poke up well above the inversions and marine layers. Intense direct sunlight and surface heating push summer afternoon temperatures into the 80s at mountain resorts, much warmer than on distant beaches trapped below their marine layer inversions. Rarified (thin) mountain air allows radiation to quickly escape after sunset, yielding cool summer nights, resulting in daily temperature swings of up to 50°F between highs and lows. Above the tree line, in the higher country's glaciated valleys near the tallest peaks, summer days may struggle to make 70°F (21°C), while any summer night can cool below freezing.

FIGURE 1-21. Summer trails into the high Sierra Nevada (such as here above Big Pine) may lead to streams and meadows fed by patches of ice and melting snow remaining from last winter's brutal deep freeze.

Despite the cold nights, bring your hat and sunscreen to guard against the fierce ultraviolet radiation that pierces through the thin mountain air. And don't forget your mosquito repellent; the irritating swarms are echoes of last winter's shrinking snowpack. Mosquitoes can become maddening just after melting snowdrifts saturate soils and leave pools of water in early summer; the buzzing should later thin out as high-country soils dry toward autumn. These warnings may sound like a foreign language to unsuspecting flatlanders from the city or to visitors who assume that all of California enjoys relatively pest-free mild and comfortable Mediterranean summers.

The summer sun glancing off of Sierra Nevada scarps can make it difficult to imagine that these mountain landscapes carved by Pleistocene glaciers will be buried under layers of bone-chilling snow within several months. Summer's dominant upper-level high pressure and sinking air masses will usually lead to months of drought and keep weather in these extreme highlands from turning dangerous. But the higher and farther north you travel, the more likely you are to discover last winter's snowdrifts still melting during July, and frost on your sleeping bag or tent on a cold August morning. These are reminders that it would be smart to hike out of the high country's abbreviated summer before you are trapped in autumn's first snowstorm.

FIGURE 1-22. Here in the glaciated Trinity Alps Wilderness of Northern California, summer meadows are lush and cool. In a few months, the annual barrage of winter storms will bury these landscapes in snow.

Just as desert summer heat kills the unprepared visitor thousands of feet below, so can dramatic and violent changes in this unforgiving, changeable mountain atmosphere soak and freeze unsuspecting visitors. Hypothermia awaits the overconfident who underestimate the power of nature in these spectacular landscapes above 10,000 feet (3,050 m). On the other extreme, dehydration surprises visitors who feel the cool temperatures and don't realize how much water can evaporate when they are breathing the thin, dry mountain air.

WHEN THUNDER AND TURBULENCE INTERRUPT SUMMER'S TRANQUILITY

We have already traveled from the seemingly endless mild summers along the California coast to the sizzling summers that never seem to end in our low deserts, up to the fleeting summers that can end at any time in California's high country. And now, summer intruders are lurking on the horizon, poised to challenge the routine. Just when we grow complacent with such regularity, nature jars us back into the constantly changing real weather world. Sudden and dramatic atmospheric disturbances in the form of afternoon thunderstorms can bring spectacular and sometimes dangerous interruptions to the calm heat and drought. Spend one summer season in transmontane California—above or beyond California's stable and milder Mediterranean maritime air masses—and you will likely experience these violent examples of how nature can swiftly redistribute and then concentrate moisture and energy.

FIGURE 1-23. Crepuscular rays radiate around a towering cumulus cloud that developed when eastern Sierra Nevada slopes were heated, sending moist air up into summer afternoon thermals.

As summer progresses, another distinctive subtropical high-pressure system migrates north out of Mexico and into the Desert Southwest, sometimes anchoring around the boundaries between Utah, Colorado, New Mexico, and Arizona, earning its name: the Four Corners High, which I mentioned earlier in this chapter. As winds churn clockwise out of this anticyclone, they often turn north along the Colorado River Valley, just before reaching California. This south-to-north wind-flow pattern will often carry sultry air up from Mexico and the warm Gulf of California (Sea of Cortez) and into Arizona during July and August.

Viewing satellite imagery, you can usually spot the boundary between this juicy tropical air flowing up from the south and east versus the relatively drier summer air flowing from the west or southwest. It may seem counterintuitive that wind flowing into California and off the Pacific from the west could be considered relatively dry. But during the summer, the cool, stable air masses over cold-water currents have low specific humidity—that is, they contain little total water vapor—compared to the monsoon air masses with their tropical connections. When that Four Corners High strengthens or drifts toward us, it can toss humid instability into the state. But when it disintegrates or moves far to the east into Texas, tropical airstreams from the southeast into the state are cut off; cooler, more stable summer air with relatively less water vapor may flow over California from the west and southwest, making way for less humid weather and a clear upper atmosphere.

FIGURE 1-24. This late-July image displays the boundary between cool, dense, stable marine air to the west (note stratus and fog along the West Coast) and very moist unstable air flowing up from the southeast, east of major mountain barriers. Towering afternoon thunderstorms are popping up over the deserts and eastern slopes of the Sierra Nevada. Southeasterly winds are pushing the thunderstorms' anvil tops toward the northwest. *Source:* NOAA/National Weather Service.

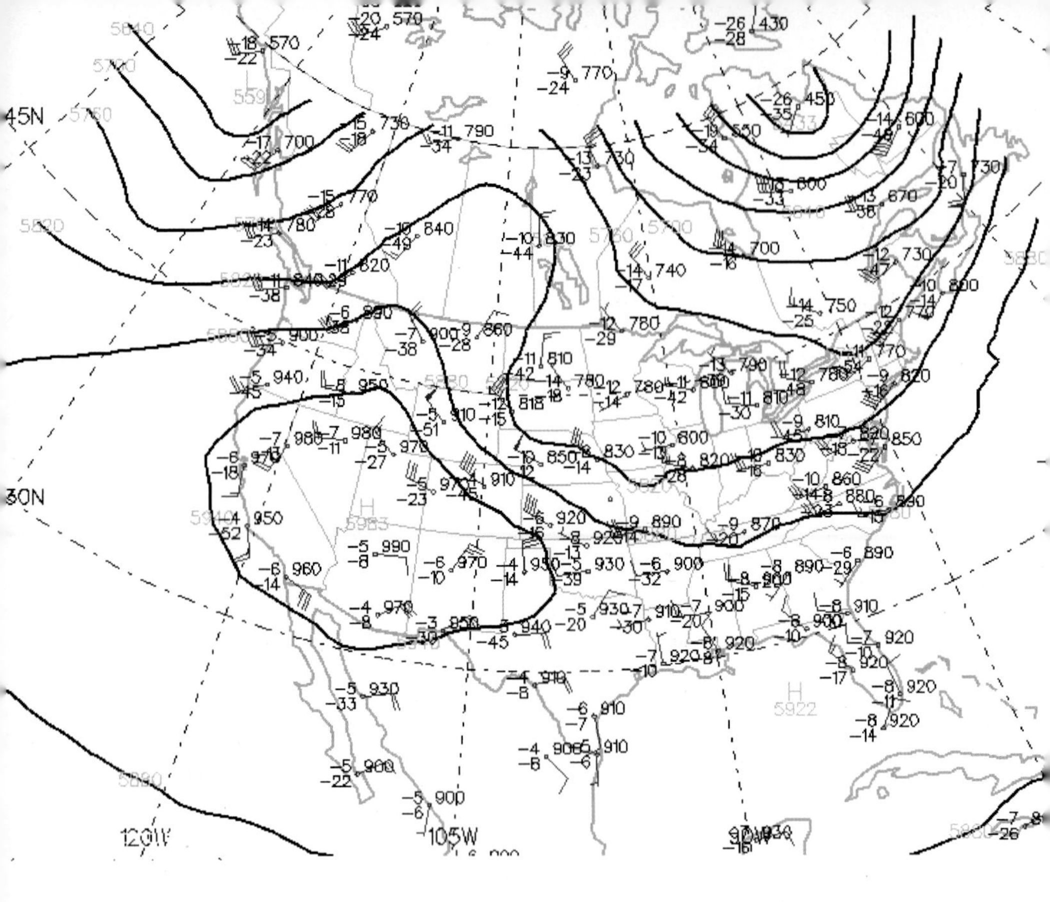

FIGURE 1-25. This weather map (at 500 mb in the middle of our atmosphere on July 10) shows how the Four Corners High pressure has expanded and drifted west of its typical summer location. Winds rotating clockwise around it are carrying moist air up from the southeast and into Southern California. Summer's more stable and relatively drier southwest flow has been pushed into Northern California and the Pacific Northwest. *Sources:* NOAA/National Weather Service and San Francisco State University.

The summer monsoon is reinforced when low-level moisture is also drawn into the Desert Southwest. As local pockets (or parcels) of air are heated during summer days, they expand, become less dense, and begin to rise. You may have noticed a miniature example of this process when a dusty whirlwind organizes on a warm, sunny day in the desert. When that warmed pocket of air expands and rises, it leaves a void near the surface; winds swirl into the center from all directions to fill the vacancy. These local dust devils may become gusty and briefly violent, but they don't create storms because the air is too dry and they have no support aloft. They do remind us how more widespread areas of thermal low pressure may form on summer afternoons and how larger air parcels can become destabilized when they are heated at the surface. The most extreme examples of this process can be found above wildfires and volcanic eruptions. Magnificent pyrocumulonimbus clouds (sometimes called cumulonimbus flammagenitus), thunderstorms, and even tornadoes can develop above these most intense heat sources, as hot-air updrafts accelerate above fires that generate their own wind and weather.

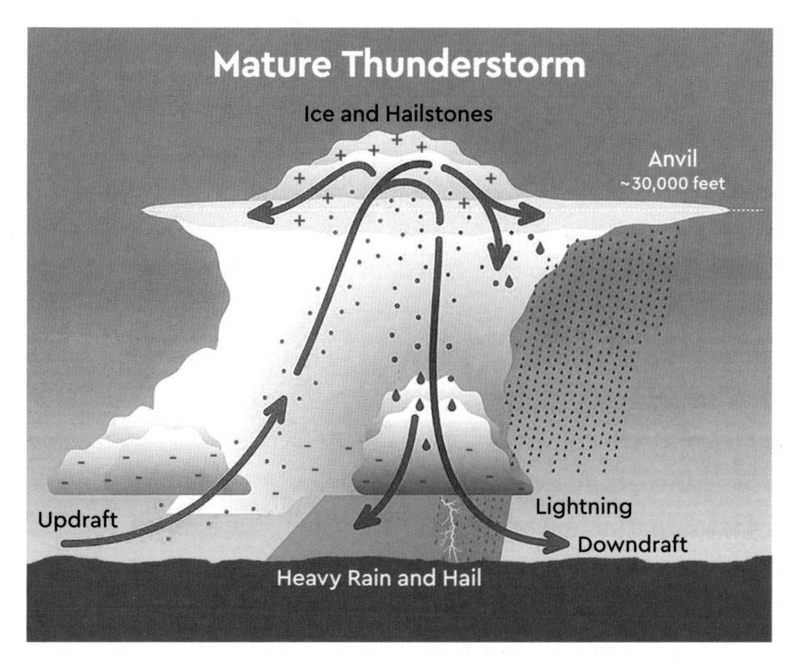

FIGURE 1-26. Towering cumulus may build into cumulonimbus and mature thunderstorms with strong updrafts and downdrafts. Such storms may briefly produce dangerous lightning, fierce winds, heavy cloudbursts, hail, and other severe weather.

TRACKING SUMMER STORMS

When tropical and subtropical moisture is injected into the Desert Southwest during intense summer heating, such moist air may rise, continue to expand, and quickly cool to its dew point. As water vapor condenses into liquid water drops, a distinct, flat cumulus cloud base appears at the saturation level. Latent heat is released when high-energy vapor condenses into liquid drops, encouraging the updraft to rise faster, forming a towering cauliflower cumulus. An afternoon thunderstorm is born as it boils into an icy, mammoth cumulonimbus cloud (thunderhead) that can soar more than 30,000 feet (9,140 m) high, an exceptionally tall cloud by California standards. The characteristic flat anvil top marks a stable layer where the updrafts finally level out. The thunderstorm may drift with the wind, carrying its violent updrafts, downdrafts, and frequent lightning. Updrafts in the storm introduce more moisture, which attaches to suspended ice crystals, causing them to grow larger. In the spirit of what goes up must come down, frozen precipitation that may include large hailstones eventually tumbles toward the bottom of the cloud. As the ice melts, it absorbs energy from its surroundings, cooling adjacent air parcels and accelerating the downdrafts. As now-cold water droplets plunge toward the surface, some of them begin to evaporate in the drier air. This next change of state from liquid water to water vapor (the reverse of cloud formation) also steals more energy from the surroundings (see figure 4-31). The chilled air dives even faster toward the surface in a powerful downburst of wind and water. Violent, cold gust fronts may be pushed ahead of the storm to scoop up adjacent warm, moist air masses, building new storms that spread out beyond the initial tempest. As these outflow winds spread

FIGURE 1-27. A summer afternoon thunderstorm builds over the mountains within Joshua Tree National Park. It will sweep across the desert, dropping violent downbursts and torrential downpours that generate dangerous flash flooding.

across the desert floor, they may also shove walls of dense dust clouds ahead of the storm. These dust storms (known around the globe as haboobs) can turn daytime into darkness as gale-force surface winds topple trees and power lines.

The biggest and most powerful storms that produce the most violent weather and heaviest rain and hail occur during periods when air masses are very wet, the land is very hot, and the upper levels of the atmosphere are exceptionally cold. Those combined wet and unstable conditions are very rare in California during summer months; the only California weather stations that might record annual precipitation peaks in summer are the farthest east, closest to the Colorado River Valley. Even there, summer precipitation accounts for only about a third of average annual rainfall. In contrast to Golden State summers, July and August are often the wettest calendar months for northern Mexico, Arizona, and New Mexico, and summer flash flooding is common east of the Colorado River. These summer weather patterns have been dubbed the Southwest monsoons, or North American monsoons. They are never as wet and reliable as the summer monsoons of South and Southeast Asia, but they are triggered by similar heating of the continent and changes in atmospheric circulation patterns.

The saguaro cactus (*Carnegiea gigantea*), icon of the Sonoran Desert, may rely on those sudden thunderstorms to survive the summer. The western edge of their natural range barely encroaches across the Colorado River Valley into a few patches of California's most eastern deserts, marking the very edge of the summer monsoon's more reliable torrents. The majestic arms of the saguaro bend upward, as if to point toward summer's thunderheads that will deliver merciful cloudbursts to this scorched earth. And saguaros are made ready for them. A four-foot-tall saguaro may have been expanding its shallow root system for 50 years until the roots radiate out more than four feet from its base. Similar to many other cacti, the saguaro has evolved to absorb whatever moisture might briefly soak into the soil. It also sends a deeper taproot to access whatever groundwater might be available.

When the Four Corners High occasionally shifts, expands, or wobbles into just the right position to push some of that summer monsoon moisture across the Colorado River Valley and into Southern California deserts, drastic weather changes may result. Middle-level clouds (between around 10,000 to 20,000 feet) may appear in summer, moving from south to north or east to west, often a sign that California is on the western periphery of that shifting Four Corners High and the boundary between dry summer air to the west and wet air to the south and east. Altocumulus clouds may decorate the sky with their tropical castles, signaling that an intrusion of moisture and instability is breaking through California's dominant dry summer high pressure. Some of the high and middle-level clouds often represent leftover moisture that has drifted from yesterday's storms that rumbled and then dissipated over the Sonoran Desert. We can only imagine the atmospheric turbulence and violence that may have set these clouds, known as debris clouds, loose. Each day's afternoon heating can provide the renewed instability required to set storm machines in motion over and over again.

Summer's first cumulus clouds usually billow above transmontane mountain slopes heated by the sun. As air rises over the mountains, more moisture is drawn in from the surroundings while afternoon clouds and storms build vertically. These thunderstorms may slop beyond higher terrain into heated desert valleys and can expand quickly as they scoop up more hot, moist air. The larger storms can produce violent and dangerous winds and frequent lightning and flash flooding that can become deadly. One location might receive a couple of inches of rain in just over an hour, nearly as much as the total annual average in these arid lands. The deluge is quickly channeled into intricately carved rills, gullies, and canyons that have been cut into weathered, exposed, and vulnerable desert materials. These implausibly sculpted artscapes, called badlands, can force the most seasoned traveler to take another route. Nearby alluvial fans and bajadas are littered with the debris flows that have stampeded out of desert canyons during severe summer storms. Particular detritus may rest for years or decades before the next perfectly positioned thunderstorm and subsequent debris flow rework the material. Meager annual precipitation and runoff are usually not consistent or powerful enough to transport the debris to the bottoms of salty desert playas. Many of the desert's geomorphic features are shaped over eons by a multitude of these very brief storms. In more recent decades, clueless humans, in their ignorance and short-term thinking, have built on these alluvial fans, only to repeatedly suffer from damaging mud and debris flows.

I've witnessed many examples of these events and landforms on display at the base of desert mountains that range from Mexico well north into the Basin and Range. One example from more than eighteen years ago comes from a visit to one of my favorite spring superbloom locations: Anza-Borrego Desert State Park. A popular campsite below Palm Canyon had been destroyed by a debris flow that raced out of the canyon the previous summer. The dead carcasses of thick, tall California fan palms (*Washingtonia filifera*, our state's only native palm) had been swept out of their canyon and into the camp that had been built on this alluvial fan. Fans piled on the fan! They were riding on top of a few feet of mud and other debris that suddenly raced and oozed through camp, squeezing through the bathrooms and encasing picnic tables.

Each year, powerful debris flows block roads and damage infrastructure on alluvial fans from Mexico into the Basin and Range. Boulders the size of cars can be rearranged as they are carried downhill on the fans. One unusually strong monsoon-style storm complex organized and then swept into Death Valley in the early morning of August 5, 2022; 1.70 inches of rain were recorded there in a few hours, breaking the all-time one-day record. Resulting floods and debris flows trapped more than a thousand visitors, even though it occurred during the park's slowest tourist season. The very next summer, on August 20, 2023, the remains of Tropical Cyclone Hilary poured 2.20 inches into Death Valley's official rain gauge at Furnace Creek, breaking their one-day record again. Although the park had been closed in anticipation of these storms, about four hundred people were stranded when roads were again washed out and destroyed.

The winners are species that have adapted to successfully compete in such hostile and occasionally violent desert environments. When sudden floodwaters recede, they may leave nutrient-rich sediment and debris after the water soaks into the ground. Some species even depend on the turbulent flows to disperse and plant their seeds. For instance, the turbulence and abrasion of flash floods and debris flows encourage the germination of smoketree (*Psorothamnus spinosus*) seeds, so you will find these trees lined up along sandy desert washes. Their roots grow quickly to tap into groundwater sources often buried below these washes, all waiting for the next desert storm. Smoketree often grows near thorny catclaw (*Senegalia greggii*), desert willow (*Chilopsis linearis*), and a few other large shrubs and small trees to provide food and shade for a variety of animals such as jackrabbits, which may become meals for coyotes and other predators. Once established, the plants function as barriers that slow and even reroute the very flash-flood currents that shape these plant communities.

During my graduate field work, I studied the spectacular alluvial fans at the base of the White Mountains, in California's Basin and Range to the north of the Mojave Desert, and far removed from those Anza-Borrego and lower-desert deposits. Local ranchers told stories about hearing the distant thunderous roar of water and debris before the source of the rumbling could be seen blasting out of nearby canyons. The flash flooding would end almost as fast as it started (in a few hours at most), leaving thick lobes of mud and debris deposits on the fans. I learned that powerful debris flows out of the White Mountains can also be triggered by unusually wet and warm winter storms (considered in chapter 3) when heavy rainfall melts layers of snow. Winter storms' destructive gooey mud baths can also spread out from canyons at the base of Southern California coastal mountain ranges, across alluvial fans, and into inland valley developments that have been creeping up to the base of steep slopes, such as along the San Gabriel Mountains. In "LA against the Mountains," a famous essay about Southern California geology, John McPhee compares the swift and violent slurries to rock porridge, bread dough, or wet concrete mixed with raisins, carrying imbedded cars and trucks, boulders larger than cars, people, and coffins. But it is the desert's isolated summer storm events that can be most surprising, sending walls of sediment-laden water, mud, and debris far beyond a thundercloud's vertical rain shafts and across desert washes. I've seen such thunderstorm events start out as beautiful, refreshing, and exhilarating, and then suddenly turn into life-threatening displays of nature's awesome power. The remains of cars and houses and the bodies of people who didn't sense the danger or heed the warnings are sometimes found downstream. Turn Around, Don't Drown!

Once the surface is cooled by the soaking downdrafts and the vertical air columns stabilize, local summer storms may march on or die out as fast as they organized. The heat sources and much of the instability usually dissipate after sunset. As calm returns, surviving residents and storm chasers are left to assess what really happened. The flood scars left in desert washes and the sweet aroma of water on otherwise xeric plants and soils may be the only evidence remaining from nature's violent summer temper tantrums. You can almost hear Gaia, in her poststorm hangover, exclaiming, "Hey, what did you think of my performance? Wasn't that awesome?"

Occasionally, this summer monsoon moisture drifts from southeast to northwest all the way up along the crest of the

Sierra Nevada, into the Basin and Range, and even into Northern California's mountains and transmontane valleys. Rare disturbances masquerading as weak renditions of winter's stronger storms may briefly weaken summer's dominant high pressure and help provide the instability necessary to trigger summer thunderstorms in the north, most commonly along mountain slopes heated during the afternoons. Those brilliant cauliflower cumulus clouds boiling over the Sierra Nevada or Cascade Mountains during summer afternoons can deliver brief downpours to the forests and high country, offering hour-long respites from summer drought. But they can also spell danger for at least two reasons. First, backpackers and other wilderness hikers may be caught off guard by summer thunderstorms with dangerous lightning strikes, high winds, hail, and cold, heavy rain that can soak the best prepared. Hikers have been killed on Sierra Nevada peaks while seeking shelter or while skidding down slippery granite slopes. Second, these lighting strikes spark thousands of wildfires each year. Recently some have spread to become the most destructive fires in California history; others have continued to burn into the fall.

"Dry lightning" is a constant threat to every California forest and a dreaded term among California firefighters. These storms can build when there is just enough moisture for condensation to set in within hot columns of air that have been rising from the simmering land all afternoon. The air finally reaches its dew point at higher altitudes that may range up to 10,000 feet (3,000 m) above the surface, where the cloud base develops. As the towering cumulus cloud grows, it generates lightning. But rain must fall many thousands of feet (from the high cloud bases) within these thunderstorms' downdrafts, often evaporating in dry air near the surface before it ever reaches the ground. Here, nature provides the cloud-to-ground sparks and gusty winds during the drought season, but little or no water that could otherwise douse the wannabe conflagrations. Trees, plants, and grass dehydrated by summer's drought make perfect kindling for sparks growing into flames.

Thus the "fire season" may begin earlier during the hottest summers, especially when mountain ecosystems are left vulnerable by years of below-average precipitation and above-average temperatures. Add dwindling soil moisture; a dry, windy heat wave; and dehydrated fuels accumulated over decades, ignited by hundreds of dry cloud-to-ground lightning strikes from a series of pop-up thunderstorms scattered across several of our mountain ranges. This is the perfect formula for overwhelming and exhausting our firefighting efforts. Such weather conditions have inflamed our fire seasons and incinerated and reshaped forest ecosystems long before the first soaking winter-style storms could come to the rescue. Summer is only the beginning of California's wildfire season; the dry tinder days of September and October—even November, if the rains are late—are the peak times for wildfires in some years.

SHOCKING THUNDERSTORM SCIENCE

Erratic by nature, summer thunderstorms have inspired countless myths and legends, particularly in our arid lands: lightning out of a clear blue sky, flash floods out of nowhere, heat lightning that somehow emerges from the ground on clear summer nights. But such awesome nature shows all originate from magnificent cumulonimbus clouds. These vertically developed towers also produce some of our most spectacular rainbows when sunlight peeks through their rain shafts. But their dazzling displays of awesome power call out for more attention here. We've seen how thunderstorms may develop when moist air masses become unstable. Such instability may be caused by heating at the surface; the introduction of cold air aloft; migrating air parcels forced to rise over mountain barriers; cold fronts scooping up relatively warm, moist air; or a combination of these and other factors. When these conditions produce extreme differences between relatively warm air at the surface and very cold air aloft, the moist air rises. As water vapor condenses and freezes in clouds, it releases the latent heat that fuels thunderstorms and other severe weather across the planet. In unstable air, saturated parcels will continue to rise higher as long as they remain warmer than the surrounding atmosphere, producing giant thunderclouds. Static electricity builds (more than 1 million volts per meter) as the top of the cloud becomes more positively charged and the bottom of the cloud becomes more negatively charged, until there is a discharge of lightning.

This powerful but narrow spark can heat the air to 54,000°F (30,000°C), causing explosive expansion of billions of air molecules and creating the shock-wave sound we call thunder. It takes the sound of thunder about 4.7 seconds to travel one mile (nearly 3 seconds to cover one kilometer). The shock waves dissipate like ripples in a pond of water until the thunder is no longer noticeable when you're far enough away. It is no wonder that there are so many different stories, originating from countless cultures, about lightning and thunder gods. In California, whether it was to acknowledge their startling power or to better understand, identify with, attract, or ward off such storms, Native Americans often incorporated sounds imitating thunder into their ceremonies and rituals. The variations of sounds and interpretations reflect the diversity of their natural environments and cultures.

There is a definite late-summer peak in thunderstorm activity over most of the state, particularly on the inland sides of the great mountain barriers of transmontane California. These places lack cool sea breezes that might otherwise stabilize the summer air out there. Thunderstorms are also most common over and near mountain slopes that have been heated by sunlight and/or represent barriers that force moving air to rise up and over them; nearly all mountainous terrain in the Golden State experiences more frequent thunderstorms compared to surrounding flatlands. Diurnally, there is a dramatic increase in average thunderstorm frequency after surfaces have been heated during the warm afternoon hours, while the fewest occur during early morning hours after surfaces have cooled.

Because thunderstorms are energized by hot, unstable air, and the coast is usually cooled by the marine layer, locations on the Pacific edge of California will experience only a few thunderstorm days each year, while that average number increases to more than a dozen as we move uphill and inland. Statewide, the frequency of cloud-to-ground strikes per year for every 2.5 acres ranges between near zero along parts of the coast to more than thirty in our higher inland terrain. According to the Western Regional Climate Center, radar measurements detect between fifty to sixty thunderstorm days somewhere in the Sierra Nevada each year. On the other hand, California is home to the four large metropolitan areas in the US averaging the fewest number of thunderstorms per year; they are all down in the flats: San Francisco at 2.9 per year, San Diego at 3.0, Sacramento at 4.5, and LA at 4.6. We will explore these places' less frequent violent storms in chapters 3 and 4.

FIGURE 1-28. Summer afternoon thunderstorms rumble over Lembert Dome around Tuolumne Meadows in Yosemite high country. Lightning and brief drenching downpours will soon spread across the meadow.

WHEN THE TROPICS INVADE CALIFORNIA

More infrequently, another more extensive wet disturbance might invade from the south and east to slice through summer's heat and drought. Late in the summer, changeable circulation patterns can give a bump to the summer monsoon by directing a tropical storm or its remnants up from Mexico and farther north, along the Gulf of California (Sea of Cortez) east of the Baja Peninsula. This slender strip of warm water provides a bit of energy that could keep these tropical surges strong enough to make it into the Desert Southwest. Once every several years, one of these dying tropical disturbances may turn in to southeastern California from the Sonoran Desert, spreading soaking rains and record-setting downpours across the deserts and inland mountains. Some of California's most historic desert flooding has been recorded during these late-summer-monsoon tropical surges. These invasions are so infrequent, desert ecosystems west of the Colorado River Valley must rely on winter storms that normally dominate meager annual precipitation totals. Still, these rarest summer gully washers can represent welcome exceptions to long summer droughts . . . until they explode into life-threatening flash-flood events. Springing from exceptional weather patterns, an even rarer rogue summer storm might drift off the Southwest monsoon and across Southern California's mountains, encroaching west into the coastal plains. A few may be rejuvenated along local wind convergence zones (where regional winds often meet, such as near Elsinore) that form over inland valleys. Otherwise, cool, stable maritime air masses will inhibit such storms from reaching the coast during several consecutive months of the summer season, a powerful signature of the state's seaside Mediterranean climate. Following the brief drama and local flooding, dry air always returns to rule through most of California's transmontane mountain and desert summer.

Rare summer rain events may also erupt when moisture from tropical storms drifts directly up the west coast of Baja, impacting all of Southern California and sometimes drifting even farther north. This may occur only once to a few times during each warm-water season. Numerous tropical storms and hurricanes commonly develop off southern Mexico in the warm ocean waters south of Cabo San Lucas and Baja California from late June through October. During these warm-water months, the atmosphere absorbs heat and moisture as tropical waters evaporate. Then the tropical air rises, cools, and condenses. This process releases copious amounts of heat that then fuels these tropical systems. Tropical cyclones organize and build only over ocean waters with temperatures near or above 80°F (27°C), so colder California waters represent tropical storm graveyards. Surfers may require wetsuits, but Californians also get to avoid having a hurricane season.

Most of these storms drench the southern Mexican coast or drift out into the Pacific Ocean with the east-to-west trade winds. They can delight surfers by sending large ocean swells up toward the Southern California Bight and other south-facing beaches, temporarily reversing our currents to briefly push water and carry sand from south to north. But the rare hurricane that might drift north and up the western Baja coast weakens over cold water, leaving only remnants to reach as far as San Diego and the Southern California Bight, where very few dying tropical storms have come ashore during more than a century. Fewer, including one marginal hurricane, have brushed the coast to do considerable damage. The California Current carries insufficient energy to keep these storms alive.

Three of these late summer storms have crossed over the Baja Peninsula to land in southeastern California with downpours and other leftovers after the winds died down. Among the few tropical systems that sailed up and along Baja's Pacific coast directly toward California before completely fizzling, another three stand out. One 1858 storm nearly made landfall in San Diego with hurricane-force winds, doing substantial damage before drifting off the coast to die. Another 1939 tropical storm hit Long Beach to cause considerable wave and wind damage, dumping more than 5 inches of rain across the LA area. You can find photos of these events at the Los Angeles Public Library. A more recent near miss (as of this writing) was Hurricane Kay, which decayed into Tropical Storm Kay, in September 2022. Its dying remnants brought gusts over 70 mph in San Diego County mountains and heavy inland downpours until it dissipated just offshore. The very next year, in August 2023, powerful Hurricane Hilary charged up the Baja coast. National Weather Service forecasters placed most of Southern California under a tropical storm warning for the first time in history. Hilary made landfall as a tropical storm just south of San Diego on August 20. Strong southerly winds that guided the cyclone also sheared the storm to pieces as it rapidly glided north over the mountains and across the California desert, heading for the Basin and Range. More than 7 inches (18 cm) of rain fell in one day on higher slopes, and a few of the highest elevations topped out above 11 inches. Some surrounding desert slopes received a year's worth of rain in one day. The flash floods, mud, and debris flows buried roads and other infrastructure and some neighborhoods, such as in the Coachella Valley. Although the coast was spared the winds, waves, and storm surge of a direct hit, record August rains totaling more than 3 inches in one day soaked many locations in the coastal plains. Future direct hits have the potential to generate high winds, flooding rains, and storm surges that could overwhelm many coastal California infrastructures, especially since they were not designed to withstand powerful tropical storms that would arrive during the traditional fair-weather drought season. One of these future cyclones could represent the disaster preparation definition of getting caught off guard. Scientists are studying how climate change might impact the frequency and strength of these storms. This research is especially important since warmer ocean waters and increasing water vapor in the atmosphere are likely to provide more fuel to power tropical cyclone engines.

FIGURE 1-29. Three named tropical systems caught in various stages of strengthening and weakening are seen in this GOES-West infrared satellite image taken on August 22, 2014. Which tropical storm do you think is best positioned to send swells and moisture up toward the California coast? Also note the inland blob of tropical thunderstorms drifting up into the Gulf of California (Sea of Cortez) toward the Sonoran Desert. *Source:* NASA/NOAA's GOES Project.

Those warm tropical waters south of Baja eventually cool during November so that southern Mexico's tropical cyclones and their swells disappear until the next summer. Their remnants eventually merge with and energize early season storms drifting off the Pacific and into California from late September into November, even directing moisture into Northern California. But before we progress into autumn, we must first rejoin our summer adventures up toward the northern edge of our state.

WASHOE ZEPHYR WIND MYSTERIES AND LEGENDS

We finally end our summer skies field trip farther north in transmontane California, looking east toward Nevada, where your hair will be lifted off the sticky nape of your neck by the Washoe Zephyr winds. These afternoon and evening winds, which often cascade from west to east down the eastern slopes of the Sierra Nevada and into Nevada, are legendary. They are especially common in late spring and summer. Such winds have gained the attention of renowned writers—from J. Ross Browne to Dan De Quille (William Wright) to Mark Twain (Samuel Clemens)—since the silver and gold mining boom of the 1860s in Comstock, Nevada, that followed California's gold rush. These peculiar diurnal wind events have challenged weather buffs and meteorologists for many decades.

Washoe refers to the traditional homelands of the Washoe people around Lake Tahoe and along the eastern slopes of the Sierra Nevada; these Native Americans lived with such winds many centuries before miners borrowed the name to describe their mining district. Today, the Washoe maintain an active presence in these lands, and their name is still used by many locals from Tahoe to Reno to describe this region. Zephyr was the Greek god of the warm west wind. Put the two together and we have a fairly accurate description of this wind. The zephyr moniker has even been used to romanticize train routes and names, including the *California Zephyr*. The first passenger train between Chicago and the Bay Area with the *Zephyr* designation ran in 1949, a name still used along that train route today. The eastbound trains pierce through scenic Sierra Nevada passes and canyons and then descend toward Nevada, as if pushed by the zephyr behind them.

Both drop west-to-east through Sierra Nevada forests across lower-elevation pinyon-juniper woodlands, and spill downhill into drier high-desert valleys dominated by Great Basin sagebrush (*Artemisia tridentata*). But why do these winds barrel down the eastern slopes of the Sierra Nevada, especially during late spring and summer afternoons when buoyant, warm air parcels should be rising along mountain slopes?

FIGURE 1-30. Lenticular cloud formations seem to cap the tufa towers at Mono Lake east of the Sierra Nevada. These clouds remain relatively stationary as stable winds form wavy patterns that blow through them on the leeward side of the mountains.

Normally, when mountain slopes are heated during the day, the air immediately above them becomes hot and therefore expands and rises, creating an upslope, or valley, breeze by afternoon. At night, mountain surfaces become colder as they rapidly radiate their heat into the thin mountain air, cooling the lowest air parcels. These colder, denser air parcels then settle downhill toward nearby valley floors as downslope, or mountain, breezes.

But the Washoe Zephyr reminds us of the many other forces that can steer our winds and interfere with these local daily cycles. On many warm summer days, these winds break the normal rules by cascading *off* the mountains and rolling eastward, *down* into adjacent valleys.

Some meteorologists have speculated that rising columns of hot afternoon air are connecting into and merging with prevailing upper-level westerly winds (winds from the west), drawing the gusts downward to the surface from aloft. But that connection, alone, has proved weak. More recent observations and research suggest that these winds howl when afternoon temperatures soar in the Great Basin east of the Sierra Nevada, causing the air in that area to expand near the surface; this creates an intense thermal low pressure there that invites winds to pour in from the west, off the Sierra Nevada, and barrel down eastern slopes, to fill this void. The results come in the form of familiar and often gusty afternoon surface winds from west to east, across high-desert valleys east of the Sierra Nevada. Washoe Zephyr winds usually subside later at night, after land surfaces have cooled and air-pressure gradients (differences) weaken.

FALLING OUT OF SUMMER AND INTO AUTUMN

From encapsulating fog and suffocating smog to sweltering heat and deadly flash floods, California's summer weather patterns are far more exciting than stereotypes might suggest. But it's time to put on your seat belts because the weather is going to get even bumpier as autumn arrives in the sparsely populated, relatively remote landscapes of northeastern California, farthest removed from popular images of the Golden State. The Modoc Plateau and surrounding highlands and valleys are some of the starkest and quietest landscapes of California—the communities there are also among the poorest in the state. As one of the least populated, most rural counties, Modoc is dominated by more traditional cultures and primary industries (such as farming and ranching). These open spaces also represent extreme exceptions to the mild California weather stereotypes—they are some of the best places to first observe and sense seasonal change, sometimes in one day! Summers are long, hot, and dusty in this isolated northeast corner of the state, where the weather can flip from oppressively hot summer to bone-chilling arctic winter before you can think about it. Seriously, if you don't pay attention to the radical changes in the weather up here, you won't survive.

FIGURE 1-31. Mt. Lassen peeks down on Manzanita Lake through whimsical cumulus clouds. A weak frontal system has passed through northeastern California, but this is barely a hint of the roaring gales, heavy precipitation, and whiteout blizzards that will visit in winter.

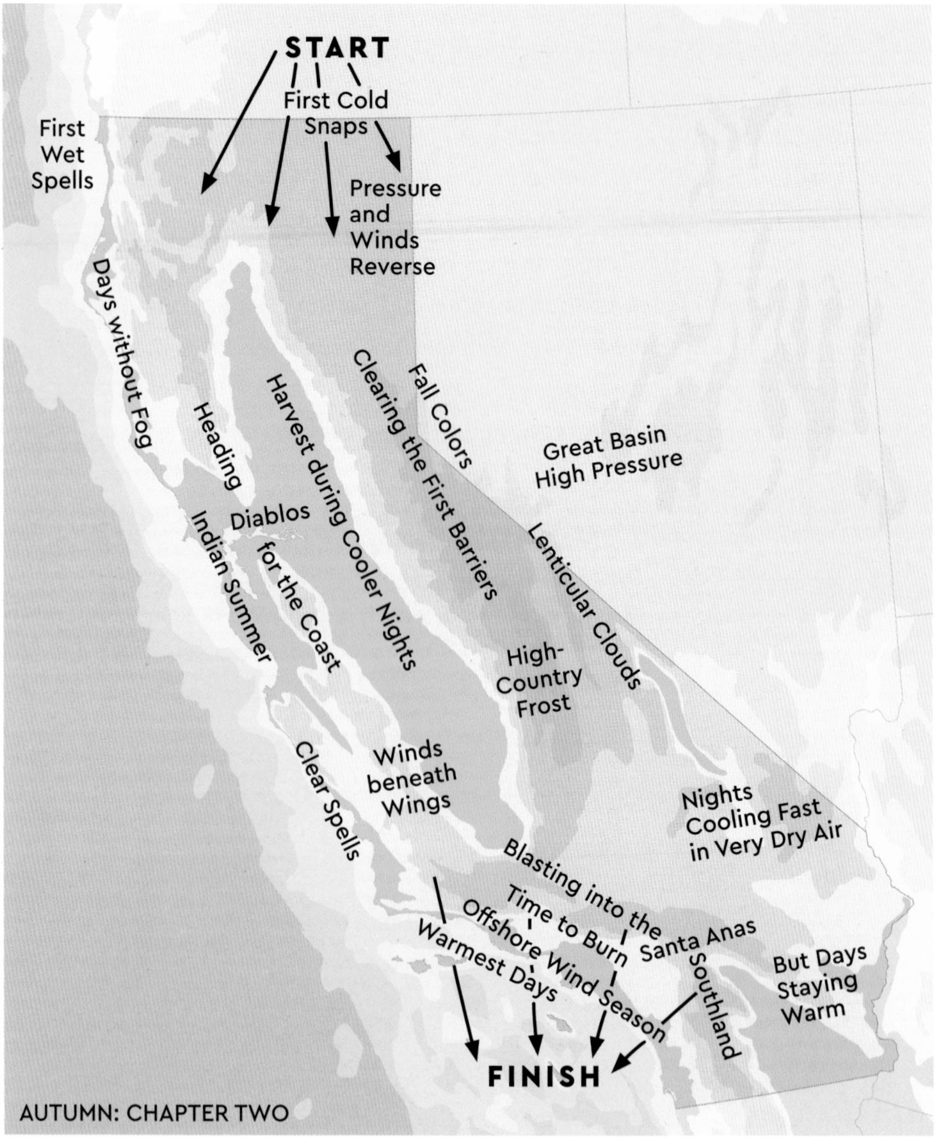

START

First Cold Snaps

First Wet Spells

Pressure and Winds Reverse

Days without Fog

Heading

Harvest during Cooler Nights

Clearing the First Barriers

Fall Colors

Great Basin High Pressure

Diablos

Indian Summer

for the Coast

Lenticular Clouds

High-Country Frost

Winds beneath Wings

Clear Spells

Nights Cooling Fast in Very Dry Air

Blasting into the Santa Anas

Time to Burn

Offshore Wind Season

Warmest Days

Southland

But Days Staying Warm

FINISH

AUTUMN: CHAPTER TWO

CHAPTER TWO

AUTUMN'S WINDS OF CHANGE

The chill of a changing season comes first to the northeastern corner of high and dry transmontane California. As September yields to October, the days grow shorter and the noon sun sinks lower toward the southern horizon. Thousands of summer stars and their constellations disappear to the west, yielding to the equally dazzling scintillations rising from the eastern horizon to announce that autumn is in the air. Just as land surfaces were first to heat up and get hottest during spring and summer, so will terra firma cool off faster and become colder during autumn and winter as compared to water surfaces and coastal locales. Summer's heat quickly radiates out to be lost through the dry continental air.

As fall becomes evident across these remote landscapes with open horizons, football shoves baseball aside. Folks split and stack firewood that will heat their wood stoves. Moods swing. *Weather wise* and *weatherize* become survival slogans. The four Ps—people, pets, pipes, and plants—will require protection from early cold snaps. And as summer winds down, it's back-to-school season.

FIGURE 2-1. Shorter days, October colors, and a chill in the air signal that summer is behind us. Deer and other wildlife around Twin Lakes in the eastern Sierra Nevada will have to migrate to lower elevations or find other ways to survive the long, snowy winters here.

LOOKING FOR SIGNS OF CHANGE

The first frosts spread from the high country to lower mountain slopes until they settle in high-desert valleys. The quick switch from summer's searing heat to autumn's biting cold in transmontane and Northern California can startle even some of the most seasoned locals. The artist Chiura Obata wrote of these autumn days, "In the evening, it gets very cold; the coyotes howl in the distance, in the mid sky the moon is arcing, all the trees are standing here and there, and it is very quiet." Patches and ribbons of high-elevation aspen and lower-elevation cottonwoods that cluster near water sources begin displaying their brilliant fall colors. Some plant species in these regions survive by going dormant through winter's hard freezes. Wildlife store

FIGURE 2-2. When yellow flowers of rabbitbrush are blooming and pinyon pine nuts are ready for harvest, autumn is arriving with its first biting cold snaps. Lenticular (lens- or saucer-shaped) clouds, sparse pinyon woodlands (which yield to juniper farther north), and high-desert scrub indicate that this is on the rain shadow side of the great barriers near the Nevada border, where summer can change to winter in a matter of hours.

up energy for hibernation or long migrations or begin simply moving to lower elevations. For thousands of years, Native Americans (from northeast California down into Washoe and Paiute regions) followed their annual fall migrations out of the high country to lower elevations, stopping where pine nuts were ripening for the autumn harvest. Yellow flowers on varieties of rabbitbrush (such as *Ericameria nauseosa*) decorate the high-desert floor, providing a source of pollen for insects during this challenging end-of-summer-drought time of year, when most flowers have long withered. This seasonal metamorphosis of environments and landscapes into autumn often spreads out of eastern Oregon and the Great Basin and into transmontane

FIGURE 2-3. You will find the remote Warner Mountains rising more than 8,000 feet (2,400 m) in the very northeastern corner of California. Patches of high-elevation white fir and even Western white pine and lodgepole pine point up toward the bluest of skies. But temperatures will soon plunge below 0°F in California's rendition of Big Sky Country, challenging even the hardiest organisms.

Northern California, down the slopes of Mt. Shasta and the Medicine Lake Highlands and into the lava beds below, across the Modoc Plateau and along the Nevada border.

One of the best places to observe summer's first waves of goodbye in the Golden State is in and around the remote Warner Mountains near California's extreme northeastern corner. At the end of long and difficult hikes, I have camped near the top of these craggy ridges not far from the Oregon border. From there I could peer down past the faulted slopes into the deep, salty high-desert valleys to see Upper Alkali Lake (and Nevada) to the east and Goose Lake to the west, which extends into Oregon. This time of year, each day's waning sunlight is a reminder of how far we've progressed past the summer solstice's late-evening twilights. After the autumnal equinox near the end of September, the nights stretch longer than the days, and these isolated mountains offer perfectly clear dark skies that reveal thousands of visible stars above. Similar seasonal swings occur all along the Oregon border, but are more noticeable in this clear, dry high country than in the gray fog on the northwest coast far to the west of the great ranges.

In Northern California's inland highlands, a hike into what was summer's lonely blessed peace and quiet can become a dangerous and deadly winter battle for survival with a dramatically quick onset. One wrong decision could be your last as the first invasions of cold air sweep across the Modoc Plateau and swirl into the southern Cascades and Basin and Range. Because so many people here work outdoors and live off the land, they are directly impacted by changing weather patterns. So these first cold snaps demand immediate attention, just when residents are anxious to see the end of the threatening fire season. While the wild animals prepare for winter, ranchers coax their cattle toward lower elevations and protection from the coming bitter winds and ice. Snow geese honk overhead, migrating from the far north toward warmer locales in Central and Southern California, along their journeys that might take them as far as the Salton Sea or Mexico.

FIGURE 2-4. Don't let this day's innocuous thin cirrostratus clouds and fair weather fool you: get ready for the deep freeze. Fall colors demand attention along the West Walker River just east of the Sierra Nevada.

The season's first sweet smells of burning firewood become more noticeable downwind of homes during cooler nights as their smoke spreads across remote villages and towns. Flat hats of smoke are sometimes trapped under increasingly shallow temperature inversions that form on long, calm nights as the coldest, densest air settles and pools near the surface. With its cold nights and the glitter of gilded aspen leaves under this terrain's big skies and open horizons, California's northeastern corner looks and acts a lot like the iconic Wild West landscapes and cultures we associate with Montana or Wyoming farther north and east. This includes harsh weather with brutal temperature extremes, which often means that these inland valleys and mountains seem worlds apart from California's southern coast, where we began in chapter 1.

With autumn's arrival, the first and most noticeable blasts of cold air are often delivered on the backsides of "inside sliders," which have nothing to do with happy-hour snacks being served in the Golden State's distant, crowded mild-weather coastal cities. These upper-level disturbances sneak into western Canada and the Pacific Northwest from the Gulf of Alaska, just as summer's dominant protector, the North Pacific High, begins to weaken and drift south. Once inland, these pockets of cold instability aloft pinch off from upper-air circulation patterns (earning the name "cutoff low"), spinning counterclockwise over land, moisture starved, but carrying their cold air along with them. The name inside slider is earned as such disturbances are routed south, on the eastern side of the Cascades and toward Nevada. They sometimes drift farther south, just east (inland) of the Sierra Nevada, spinning cold air and destabilizing the atmosphere as they progress. Some meteorologists have labeled these troublemakers "Tonopah lows" when they spin around western Nevada.

The result is a startling change from hot summer to cold autumn weather in transmontane California, all the way down to the Owens Valley. Although they don't usually carry the moisture we would expect from winter storms that stream in directly off the Pacific, these inside sliders are unstable enough to bring the first cold winds and dustings of snow to inland high country. Should they stick, the first blankets of high-elevation snow can mean an abrupt end to warm weather in any particular microclimate. The bright powder will reflect more radiation; such high surface albedo (the proportion of sunlight reflected by a surface) prevents the absorption of warming sunlight, resulting in lower temperatures and an early start to winter. Summer's hot, dry continental air masses that roasted the landscape are suddenly replaced by winter's cold, dry continental polar air masses.

Decades ago, when I first began teaching at Santa Monica College, I led my first four-day natural science field trip to the Owens Valley and the White Mountains Research Center one early November weekend. One of our first stops was at the old National Weather Service office in Bishop, now closed. While students learned how meteorologists were preparing the local forecast, I turned my attention to the weather charts. They showed a deep upper-level low dropping straight south from Washington and Oregon into western Nevada. It was a classic inside slider, promising bitter cold and some snow in a day or two. I quickly adjusted our itinerary so that we weren't caught in the advancing deep freeze. Even so, on our last morning at the field station east of Bishop, thirty-plus students and I awakened to a few inches of snow on the valley floor, which lies at just over 4,000 feet (1,220 m) elevation. The inside slider had passed through overnight, leaving a perfectly quiet, crystalline white blanket stretching from the Sierra Nevada on the western side to the top of the White Mountains on the east (both with peaks over 14,000 feet [4,267 m]). Students found themselves in a perfect winter wonderland in autumn; it was the first time some of these coastal flatlanders had experienced the swift arrival of nature's icy grip.

On another November night the next year in the same region, our field science and natural history students were huddled in tents as the cold wind dropped the temperature to 10°F (–12°C), forcing everyone to learn quick, common-sense lessons about how to avoid hypothermia and frostbite. Even during trips taken in late September, we've wakened to a first light dusting of snow adorning steep eastern Sierra Nevada escarpments to our west.

FIGURE 2-5. Glacial moraines meander and spill out across eastern Sierra Nevada fault scarps north of Bishop, reminding us that much colder and wetter Ice Age climates once ruled here. The colors tell us that after at least four months without snow, this season's first high-country flurries may be just days away.

This transitional time of year is also when whimsical wafts of steam fog condense over mountain lakes. The lakes remain relatively warm with summer's stored heat, and this warmer water evaporates into the colder air arriving above. Since cold air has a lower capacity to hold water vapor, the air quickly becomes saturated to form thin layers of fairy-tale fog above lake surfaces. This is similar to the steam fog that twirls over a mug of hot chocolate on a cold night. It is sometimes called arctic sea smoke. Steam fog also rises over hot springs in volcanic areas, such as Lassen Volcanic National Park and The Geysers geothermal field near Clear Lake.

One area you might not expect to find hot springs is on a ski slope—at least I didn't. Many years ago, while skiing down

FIGURE 2-6. Early morning sun illuminates steam fog over Tioga Lake in the high Sierra Nevada. At around 10,000 feet above sea level, the lake warms into late summer. Now the mornings are getting colder, so water evaporates from the lake surface to temporarily saturate the chilly air above. Winter can't be far away.

Mammoth Mountain alone through the quiet fresh powder, I slipped and took yet another spill. The world seemed to stop. I was lying spread-eagled within the preciously pristine silent calm of that moment while giant fluffy snowflakes fell on my face. I noticed an odd sulfur smell, so I turned around to see puffs of steam fog rising from a nearby hydrothermal vent that had melted a hole though the snowpack. It was another

FIGURE 2-7. An early season (Oct. 11) inside slider spins over Nevada. Cold, clear, stable north winds flow down coastal California and then turn counterclockwise, spinning a coastal eddy into the Southern California Bight. The winds continue turning back toward the north over Arizona and Utah. Closer to the low-pressure area, unstable cold showers and scattered snow squalls are forming from the crests of the Sierra Nevada and east across Nevada. *Source:* NOAA/ National Weather Service.

perfect example of what happens when moist, hot air quickly cools to its dew point.

Lake-effect clouds and even rain also occur above and downwind of larger water bodies, such as Lake Tahoe. Cold air masses moving over warmer lake surfaces can create instability that encourages moist air to rise and expand above the lake (where evaporation from the lake surface adds more moisture) until the air cools to its dew point and condenses into clouds and precipitation leeward (on the downwind sides) of these lakes. Rare waterspouts may even touch down within the wetter, most unstable air masses as local storms develop and grow. Greater differences between the warmer lake surfaces and colder air aloft will cause greater instability. All of these are small-scale disturbances, however, and local topography (especially surrounding Lake Tahoe) plays a much greater role in determining annual precipitation totals and microclimates in the Golden State.

PRESSURE AND WIND PATTERNS REVERSE

Such early invasions of cold air often settle over the Basin and Range and into the Great Basin during autumn. And while inland temperatures begin to plummet, coastal waters may still be near their annual warmest, thanks to a long summer of accumulated heat that has been absorbed and efficiently stored in the ocean. As the Great Basin and Southwest deserts cool, milder coastal locations may record their highest air temperatures of the year. These reduced temperature gradients, following on the heels of summer, can cause pressure gradients to ease and even reverse during autumn; this ends the consecutive months of dominant onshore sea breezes and ushers in the season of occasional offshore land breezes. Gradual changes are punctuated by sudden upheavals in the atmospheric circulation that was so stubbornly stagnant during summer.

CLEARING THE GREAT BARRIERS, HEADING FOR THE COAST

Great Basin high pressure builds in the colder dense continental air, replacing the hot thermal low pressure that dominated summer weather from the Basin and Range down into southwestern US deserts. As winds circulate clockwise out of the inland high pressure (anticyclone), they encounter massive mountain barriers (from Oregon to the Mexican border) before they can drain toward lower pressure near the still relatively warm coast. The heavy, dense air is first pushed up the dry inland sides of the highest mountain ranges as pressure builds in the Great Basin, similar to how a boulder might be rolled toward the top of a mountain. Searching for a path of least resistance over these rock walls, the wind is then channeled and funneled through canyons. As these air parcels head toward the coast, they will finally roll over the crests (such as the Sierra Nevada, Coast, Transverse, and Peninsular Ranges) and squeeze through lower mountain passes. This creates what scientists call a venturi effect, similar to how water is channeled through the nozzle of a garden hose or sprinkler and is then suddenly free to blast out of the constricted end. Once past their barriers, the heavy, dense, stable air masses are free to tumble down cismontane mountain slopes, into adjacent coastal valleys, toward sea level. The cascading, sinking air masses steadily lose their cooler characteristics as they descend. Although they may originate in cool autumn and winter continental air masses, they are heated by compression as they cascade down mountain slopes and toward coastal valleys. So, what started as cool continental air quickly warms as it descends toward the coast, and when the winds pass over warmed surfaces during the day, they will accumulate even more heat.

Each of these autumn wind affairs evolves and behaves differently, but there are some common patterns. Because the cold air first sweeps down from the north before settling on the east side of the great mountain barriers and the Great Basin, Northern California is usually the first to feel the change. As high pressure drifts east, often from the Pacific Northwest, the winds may first appear as "North winds" that descend directly over Redding and scour out the northern Sacramento Valley to paint crystal-blue skies with unlimited visibility. As migrating cold high pressure drifts farther inland to settle in the Great Basin, northerly winds gradually shift to blow from the northeast across the state (often the peak of the event) until they become more easterly over a few days. This often signals that the Great Basin High will soon drift far enough south and east to be out of the Golden State's weather sphere of influence. We will have to wait for the next offshore event (delivering a new cycle of gradually shifting north-to-northeast-to-east winds) to evolve. Intervals separating each offshore wind performance could span a matter of days or many weeks, depending on fall and winter weather patterns.

But as these pressure and wind trends migrate south down the state, they may not be strong enough to scoop out the dense, stagnant haze in other parts of the Central Valley. Instead, they often cascade down western Sierra Nevada slopes (sometimes called "Mono winds," as they have appeared to flow west from the Mono Lake region) and then over the low-level valley inversion on their way to the coastal mountains. More powerful Mono winds have battered down huge trees as they blasted through mountain valleys such as in Yosemite. But even these stronger winds may skim over shallow layers of haze and winter's radiation fogs that are often trapped at the bottom of the state's great valleys. As the winds continue west into and across the Coast Ranges, they must again push over the crests and funnel through passes and canyons, until they finally arrive at lower elevations near the coast.

In the Bay Area, these hot offshore winds are named after the Diablo Range and its Mt. Diablo, which rises between the Central Valley's Delta and the East Bay. These dry Diablo winds flow from the continent toward the Bay and coastal cities, a complete reversal of the sea breezes that dominated throughout the preceding summer months. They have sometimes been named Santa Lucia winds for the local range of mountains in Central California's San Luis Obispo and Monterey Counties; record high temperatures (over 100°F [38°C]) with extraordinarily low humidity have been recorded along those cherished coastlines when such winds tumble over the crests and rush out to sea. Some researchers have recently proposed that all of these similar Northern California offshore wind episodes be grouped together and combined under the label Diablo-North winds.

Every coastal Californian can remember abrupt swings from damp cold to dry heat brought by these offshore events. I recall one October evening when I lived in Northern California, driving with a friend near Muir Woods and along the undulating slopes of Mt. Tamalpais north of San Francisco. A hot, dry breeze from the east boosted temperatures up to 80°F along more exposed slopes, as if the desert had come calling in the forest. It created a sense of foreboding in this environment famous for its moisture and dense greenery, as if something had gone, or was going, terribly wrong. Each time we dipped into a protected canyon, we discovered how the cool, moist air more common to the forest pooled in isolated low spots, until we were thrust back up into the eye-watering wind. Most folks think of San Francisco nights as fog chilled, but in the autumn, as Eric Burdon and the Animals sang, residents love their "warm San Franciscan nights." The City's old-timers and all those who have grown up around the Bay are familiar with these seasonal cycles; visitors perceive them as bizarre aberrations. They are spectacular, precious autumn days when folks in the San Francisco Bay Area have long celebrated the fleeting warmth they didn't get all summer long.

FIGURE 2-8. There is likely to be an offshore flow when you can see so clearly from the Berkeley Hills over UC Berkeley, across the Bay, and into San Francisco before winter's turbulent storms arrive. The battle between fleeting warm, dry offshore winds and the return of the cool, moist sea breeze is on.

Autumn offers a small window of opportunity to catch some warm sun along the coast north of San Francisco, all the way to the Oregon border, as brief offshore trends sweep away summer's persistent fog, and winter storms have yet to bring their drenching rains. It can be a time when brilliant weather along the northwest coast briefly transforms the landscapes and cultures so often defined by abundant moisture. However, in the aftermath of historic droughts that have plagued California in the last twenty years, these warm winds whip down trees and power lines onto dry grass; they also bring a foreboding about dry-season fires igniting on mountain slopes. Locals strap on their masks as smoke pours out over the Bay and northern coast like the bad hot-weather twin of the usual fog.

BLASTING INTO THE SOUTHLAND

The first Europeans to arrive in Southern California recorded how the warm, dry offshore winds of autumn and winter blasted out of mountain passes and spilled into the coastal plain. Although they affect a much larger region, they have been labeled Santa Ana winds. Lively etymology debates about the origins of this name all lead back to the Santa Ana Mountains and Canyon. By contrast, there is no debate over what causes these winds. By the time the hot, dense, and sinking air reaches near sea level, autumn daytime high temperatures have often exceeded 100°F (38°C), with relative humidity driven well below 10 percent. When strong upper-level high pressure teams up with these offshore surface winds (see figure 2-9), it results in some of the hottest temperatures ever recorded on the coastal plains . . . in September and October, when the rest of the continent is cooling off! As these autumn cycles often continue on and off into winter, Santa Ana winds can warm the air well into the 80s F into January in Southern California, producing those clear, dry days that LA's entrepreneurs have long touted to attract millions of people from colder climates. Where the Santa Anas continue to blow, temperatures in breezy parts of the coastal plain may remain in the 70s through some autumn and winter nights. Meanwhile, adjacent calm, wind-sheltered locations, where cool, dry nighttime air settles without direct mixing and compressional heating, can be 20°F colder. Millions of Californians look forward to the warm autumn evening breezes. But Joan Didion spoke for others unnerved by the hot breath of the Santa Anas when she wrote, "It is the season of suicide and divorce and prickly dread, wherever the wind blows."

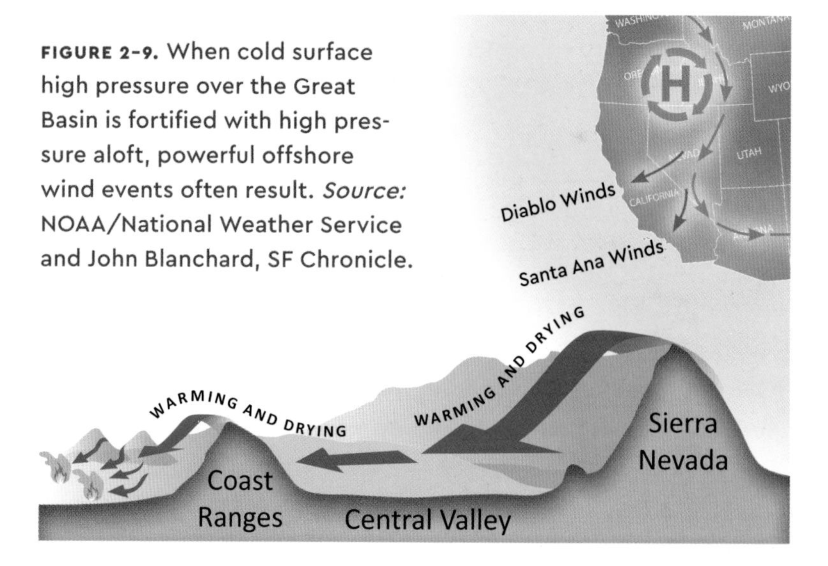

FIGURE 2-9. When cold surface high pressure over the Great Basin is fortified with high pressure aloft, powerful offshore wind events often result. *Source:* NOAA/National Weather Service and John Blanchard, SF Chronicle.

When high pressure in the air behind the mountains strengthens further, the Santa Ana gales blast through Southern California's mountain canyons and passes (such as Tejon, Cajon, and San Gorgonio) and sweep through coastal valleys with a vengeance. Any remnants of marine layers may be blown hundreds of miles out to sea. These memorable windstorms usually blow when upper-level high-pressure ridges drift over the state and team up with the Great Basin High, enhancing offshore surface pressure gradients that trend from land to sea during fall and winter months. As they shift and evolve to explode out of one aligned canyon and then another, they can resemble those giant Jurassic Park dinosaurs, stomping through neighborhoods and whipping their tails back and forth, leaving paths of destruction behind. Inland Empire locales aligned below Cajon Pass are particularly vulnerable to the beast: San Bernardino, Riverside, Fontana . . . like those Jurassic Park movies, there will be plenty of future sequels to this huffing and puffing madness.

More often, gentler offshore winds drain through mountain canyons and passes, clearing the air until they might meet a shallow sea breeze and marine layer. This dramatic boundary between air masses near the surface is usually visible because a region's milky haze and brown air pollution accumulate here, piled on the seaward side against crystal-clear air on the inland side of the front, where the offshore breezes are blowing. On many days, the warm, dry offshore flow skims up and over the cooler, moist, heavy marine air. The "hydraulic jump" is another term used to explain this wind behavior. Just as high-velocity river water must bounce up and over a large boulder in its path, so might these offshore winds spring up and skim over the shallow, dense, stubborn marine layer front where the two meet. On these occasions, summer conditions briefly extend into fall and winter at locations just a few miles inland from, or a few hundred feet above, the front and inversion. The difference could be more than 20°F between the cool marine air and the hot desert-like air with unlimited visibility. Because predicting exactly where this boundary will migrate and oscillate back and forth is almost impossible, weather forecasters face a major challenge when they are expected to tell us whether our neighborhood will experience a moist, hazy, 65° day or a crystal-clear, dry, 90° day. Such weather extremes might be found only a few miles apart. Malibu's misty beach has experienced late-autumn afternoon temperatures of 57°F (14°C) while just 10 miles away, on the inland side of the chaparral-covered mountains, neighborhoods bask in dry offshore breezes heated to 88°F (31°C). It's a classic autumnal personality disorder: premature winter on one side, stubborn late summer on the other—and the boundary between is always shifting.

Once the winds blast out of the canyons into inland valleys and then across coastal plains, there is little uniformity. Motionless patches of invisible air will settle within narrow sheltered paths, or local eddies may circulate or even swirl in opposite directions nearby. Gales howl out of a canyon aligned with the offshore wind, generating ruffles and ripples over the ocean's surface. They carry over the water until they spin surface eddies that may even locally swirl the air currents back toward the land. Scientists have confirmed what seasoned sailors, hang gliders, and even bikers who regularly traverse the rugged coastline on Pacific Coast Highway know: calm winds in the shelter of a protective hill or coastal cliff can become violent gusts as you pass into a perfectly aligned canyon. Or you may find calm winds in canyons that are carved in opposite directions from the wind while those offshore winds howl over the mountain crests. Such localized playful devil gusts perform as mischievous magicians, blasting you off your feet here and disappearing there.

Tree deformation studies have been conducted to determine where the strongest and most damaging winds are more common. For instance, one such study was done to assess the direction and intensity of Santa Ana winds around and through Newhall Pass north of Los Angeles. You can conduct your own informal study by looking for signs as you move through your neighborhood. As an example, you would expect all of those introduced iconic tall, skinny palms along the Southern California coast to lean inland, since the sea breeze dominates for most of the year. You may instead notice that they lean toward the ocean in many neighborhoods, signaling that strong, dry, desiccating Santa Ana winds occasionally sweep through those locations.

FIGURE 2-10. This Tyrannosaurus seems to be annoyed by the gusty offshore Santa Ana winds as they funnel toward the coast through San Gorgonio (Banning) Pass. Surrounding fan palms (they are real) also seem to be struggling to resist deformity.

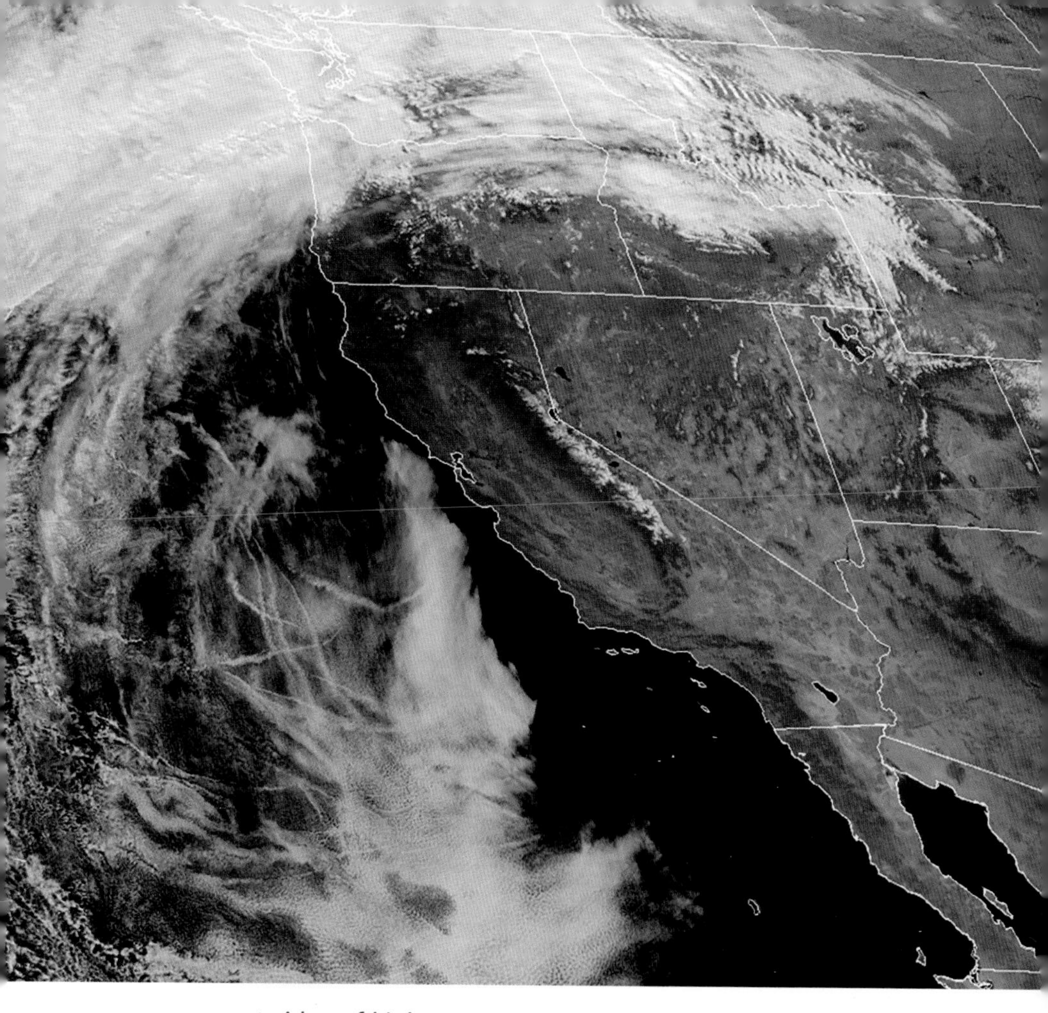

FIGURE 2-11. A ridge of high pressure creates a warm, dry, offshore flow over California. Note that storms are being directed far to the north into Washington and that there is an early dusting of snow covering Sierra Nevada crests. The marine layer has been pushed out to sea during this late October dry spell. *Source:* NOAA/National Weather Service.

North winds usually first burst through the passes of the Transverse Ranges in Santa Barbara and Ventura Counties and those of northern LA County. San Gorgonio and Santa Ana Canyons often become the focus as these winds sweep south and turn more easterly. Winds that blast through the Inland Empire and Santa Ana Canyon into Orange County have been known to bump temperatures up more than 20°F in an hour as the shallow marine layer is suddenly swept out to sea. As a kid, I remember waiting for the Santa Ana wind in the cool, moist marine air near the surface and hearing the first rustling of leaves high up in our towering eucalyptus tree. Within minutes, eddies of hot, dry offshore winds swirled downward toward the surface, scouring out the moist air, announcing the arrival of the latest heat wave driven by Santa Ana winds. Here was this weather kid growing up in Santa Ana studying Santa Ana winds: happenstance?

Randy Newman's classic song "I Love L.A." includes a line most Southern Californians understand, though it often puzzles folks from other parts of North America: "Santa Ana winds blowing hot from the north." How can winds from the north be hot? As we've discussed, they may have started out cool, but they are heated by compression while descending to lower elevations on coastal slopes of the Transverse Ranges and into the LA Basin. During this seasonal switch of wind directions, the coast becomes a temporary desert and the Transverse Ranges become gigantic and very efficient barriers to any invasions of cold air masses from the north. These crosswise mountain ranges trending west–east from the Santa Ynez to the lofty San Bernardino Mountains look down on coastal regions from Santa Barbara through the LA Basin and Inland Empire. It's as if Earth and atmosphere were involved in some kind of conflict: I'll build massive mountain ranges to block your attempts to introduce northern winters into Southern California's coastal plains.

Farther south, from Orange and Riverside Counties to the Mexican border, the Peninsular Ranges, which trend north–south, provide a barrier to invasions of cold fall and winter continental air masses from the east, just as they had blocked sea breezes and their marine air masses from getting into the deserts during summer. Following patterns dictated by the laws of physics, fall and winter offshore winds must tumble downslope from east to west into coastal valleys, only to be heated by compression. San Diego County population centers feel the dry autumn and winter heat on the coastal side. Although such winds may spread south a little later with a little less drama, they are sometimes whipped faster by pesky low-pressure systems spinning offshore, pulling the breezes off the land. And as with any region on the coastal sides of Southern California's mountains, San Diego County is subject to heat and fire danger. Although the specific numbers and details are always somewhat different, most seasoned Californians have heard the story before: when autumn temperatures soar above 100°F (38°C), and the compressed winds drive the humidity below 10 percent before seasonal rains have had a chance to wet the soil, and fuel moisture is low, it's red flag warning time.

And here we are, back where we started our weather journeys through California, trapped in traffic in those crowded coastal cities or walking in the sand along Southern California's archetypal beaches. We have returned to where the sunny, endless summers seem to stretch on forever . . . at least until winter's first storms finally arrive. We are reminded how our weather, time, and seasons wait for no one, and why Joni Mitchell wrote and sang about the seasons going round and round . . . like painted ponies, we're captive on the carousel of time.

Each offshore air cascade wanes as high pressure drifts farther away to the east, causing pressure gradients to lessen until the next cycle. As the gusts settle down, there are more variations in wind direction within the day as well. Onshore breezes spread back toward the land during each afternoon after land surfaces are sufficiently heated and the warmed air over the continent expands a bit. Offshore breezes drain back out to sea,

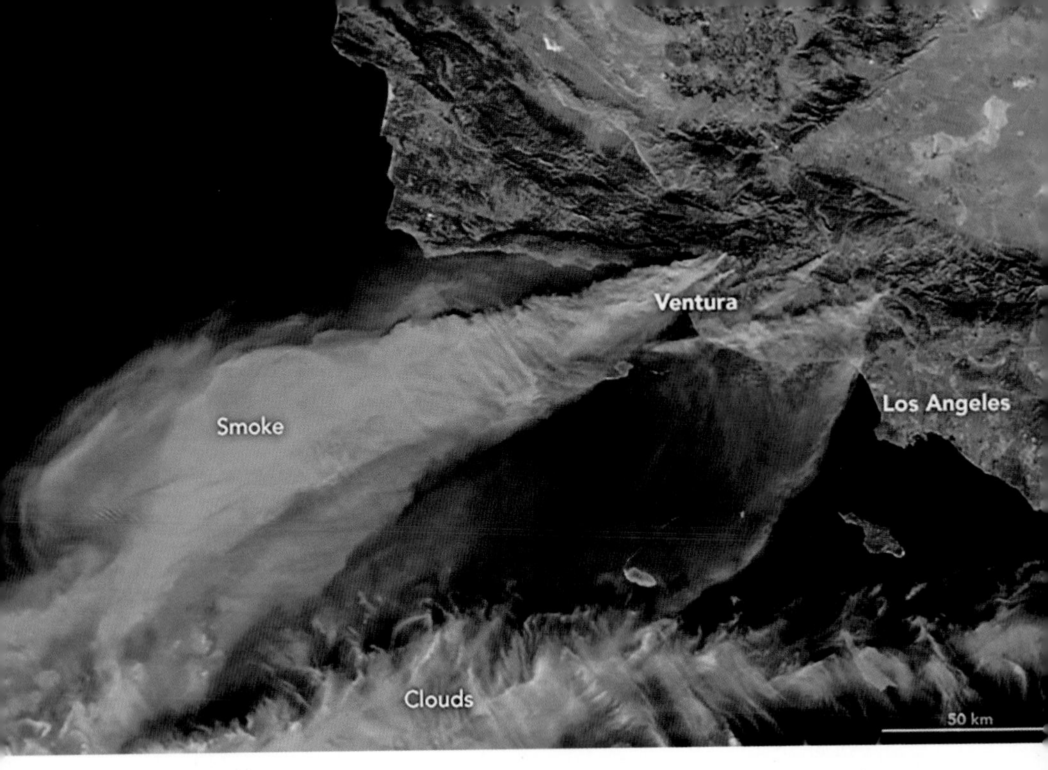

FIGURE 2-12. This image from space shows three major wildfires burning northwest of Los Angeles on December 5, 2017. Hot, dry Santa Ana winds from the northeast are blowing the smoke far out to sea. A stream of high, thin cirrus clouds to the south contrasts with, and is not related to, the smoke. *Source:* NASA Earth Observatory Satellite.

dominating during night and early mornings, after land surfaces have had a chance to cool off, building dense, high pressure near the surface. This cycle of sea breeze by day and land breeze by night is commonly recognized in many locations around the globe and has been familiar to mountain and island Indigenous people for centuries, from the Pacific Islands to California and beyond. Understanding the complex interactions between local geography and air pressure helps us see why California's onshore breezes remain dominant throughout summer afternoons, and why offshore breezes are common during autumn and winter, especially later at night and in the early mornings, to surfers' dawn-patrol delight.

CELEBRATING AUTUMN, PREPARING FOR WINTER

Wildlife and humans in the northeastern corner of the state get started on their winter prep early. But as the days shorten and cool, animals and plants throughout California begin to plan ahead for winter's rains and snows. Acorns ripen for harvest on many oak tree species during autumn months; such nourishing morsels support entire ecosystems. Most folks have seen squirrels scampering around and fighting with birds and other animals for acorns, and you might have heard the laughing call, drumming, and tapping of acorn woodpeckers (*Melanerpes formicivorus*). These colorful red-capped birds store their acorns in relatively soft tree bark holes (such as in ponderosa pines) and telephone poles to establish their own granaries. Recent research suggests that these animals (especially squirrels and jays) have better recollection of their storage sites—a skill that comes in handy during leaner times—than we previously thought.

Oak woodlands are especially productive throughout extensive regions of cismontane California. Grizzly bears (*Ursus arctos*) once foraged by the thousands amid nature's autumn buffets before Californians exterminated them in the early 1900s. Smaller and more docile black bears (*Ursus americanus*)—not all of which are black—have rebounded despite ongoing hunting and are now so numerous throughout the state (up to thirty thousand individuals) that they wander into foothill suburbs, sometimes when stressed by extreme summer heat and drought. These resilient omnivores are especially common in cooler forests from the North Coast Ranges to the Cascades and Sierra Nevada. Although protected with thick fur, many black bears practice a modified form of hibernation (called torpor, or just dozing off for weeks or months) to escape tough times during colder winters in higher country. In preparation for cold winters, they forage up to twenty hours each day until they have increased their weight by one-third and added layers of fat up to four inches, while the insulating value of their fur can double.

While headlines might gasp about backyard bears, my own California black bear encounters have all occurred in or just above woodlands and forested mountain terrain. They have appeared as startling surprises along wilderness trails and as invaders into camp, including during field trips with students. We were able to chase off most of them; others were more stubborn and not to be challenged.

Bears are one of the many animals that benefit from the dense nutrition of acorns—they can eat up to eleven pounds of acorns and consume 20,000 calories in a day. On one long October weekend in Kings Canyon National Park, returning to

FIGURE 2-13. Don't disturb this hungry black bear while she's harvesting acorns in October in Kings Canyon.

camp in our packed van with a trailing caravan of students, we spotted a black bear climbing along the limb of an oak tree that hung over the road! We stopped to observe her behavior and noticed that she was slapping at the tree's ripe hanging acorns. She consumed some of them in the tree and knocked others to the ground. She eventually scampered out of the tree and on to the road and ate the acorns she had knocked down, forcing a few amazed passersby to swerve to avoid her. She was driven to accumulate as many calories and other nutrients as possible before the first snowfall, and she wasn't going to allow humans to interfere with her survival instincts. One weekend later, that entire section of the park closed for the winter, eventually locked in by snow. It left us wondering whether she made it through the long, snowy winter after her intense acorn feast, which we hoped had set her up for her seasonal slumber.

A TIME TO BURN

The timing, nature, and strength of autumn's offshore winds, and the ignorance of settlers who didn't recognize California's predictable patterns, have combined to produce the greatest wildfire disasters in the history of the state and the nation. Winter's and spring's rains encouraged new growth, followed by the long, dehydrating drought of summer, followed by some of the highest temperatures, lowest humidities, and highest winds of the year. Dried-out grass, leaves, and twigs abound in wooded canyons and brushy hillsides. They are all powder kegs waiting for sparks. We might echo Annie Dillard, who wrote of her local fire-prone landscape in the Southeast, "Everywhere I look I see fire; that which isn't flint is tinder, and the whole world sparks and flames." As autumn progresses, huge swaths of inland mountain forests, especially in Northern California, have already been incinerated following summer's intense drought and heat waves. And now, as autumn's scorching blowtorch winds sweep off the continent, it is time for the coastal mountains to burn. Their coastal sage scrub, chaparral, grassland, and woodland plant communities are well adapted to the annual fire cycle. And once the settlements spread into such wildlands, it was as if nature were playing with humans in a deadly gyrating pyromaniac dance. You can feel the landscapes and communities anticipating (or dreading) the ignitions that will create the latest conflagration that will rage through what has been coined the wildland–urban interface, where oblivious people expanding their encroaching developments must eventually look nature in the eye. In this Mediterranean environment with increasing human imprints, you couldn't write a science fiction script that could produce more perfectly dangerous and terrifying fire and flood seasons. Talk to anyone who has experienced or fought these fires fanned by California's offshore winds. You might be able to steer the flames, but you can't stop them.

The spark could come from target shooters, a chain scraping on the pavement, a lawn mower or other landscape equipment, a neglected campfire, aging or damaged electrical equipment, an arsonist, or even the occasional "dry" lightning (a bolt from a thunderhead carrying more lightning and less rain). While those millions of us living in unmovable shelters built from dried wood experience fire as a threat, the coastal plant communities that help define Mediterranean ecosystems have evolved to require intermittent fires for their species' very survival; many of these plants have even evolved to encourage the flares, so that their seeds and burls can sprout. Pyrophyte endemics (such as wildflower species

FIGURE 2-14. Strong Santa Ana winds pushed this November 2018 Woolsey Fire over the Santa Monica Mountains (right to left), through tinder-dry plant communities and then directly into and through Malibu neighborhoods. The flames finally stopped at the beach. These firestorms can generate towering pyrocumulonimbus (cumulonimbus flammagenitus) clouds and their own local weather patterns.

that include some lupine and phacelia) germinate in the aftermath of fires, while common chaparral and coastal sage shrubs crown sprout (regrow from their root crowns).

Recent scientific studies that include analyses of tree rings (dendrochronology), ancient charcoal deposits, and weather records reveal what all of us suspected: the recent wildfire catastrophes in California and the West are unprecedented. Through most of the last century, people often attempted to extinguish every wildfire that erupted, and we are now paying the price. The last few years and decades have hosted the largest, most deadly and destructive wildfires with the greatest areas burned. Annual acreage burned each year increased by several

FIGURE 2-15. Calamitous firestorms burned through neighborhoods north of the Bay Area in the autumn of 2017, destroying more than eight thousand homes and businesses (including entire shopping centers) and killing more than forty people. Here in northern Santa Rosa, walls of flames raced down the hills (in the background), fanned by Diablo winds that gusted up to 70 mph. The least fortunate didn't make it out alive.

hundred percent in California's chaparral and scrublands from the 1980s to 2021 and by well more than one thousand times in our forests. This is a result of a synergy of record drought and climate change; both clear-cutting the big trees that can survive fire and suppressing natural fires that would slowly clear out brush and smaller trees; the introduction of invasive nonnative species that dehydrate into massive stores of low-moisture fuels; and more people moving into the wildland–urban interface. For instance, the introduction of invasive and now ubiquitous cheatgrass (*Bromus tectorum*) from Eurasia has created tons of dried kindling, which serve as convenient conduits for spreading wildfires, even throughout our deserts. Survey California slopes, especially grasslands and woodlands, and you are likely to notice this plant sprouting and growing in winter and spring. But it quickly withers to crispy stalks and sticks, as though it were wondering how it got stuck in this foreign land of intense summer sun and prolonged Mediterranean drought. As the type and intensity of precipitation has been changing and severe heat waves have become more common, summer soil moisture levels have been measured at historic lows, even when considering the paleoclimate record. Every year when fires get bigger and fire seasons last longer, the same words echo across the landscape from firefighters throughout the state: "This is unprecedented; we've never seen wildfires behave like this before." The many

FIGURE 2-16. A fire blazed through Zuma Canyon in Malibu, burning the coastal sage scrub to the ground. No problem. It provided a fresh opportunity for several species of fire-adapted flowering plants to sprout and make a remarkable recovery. This photo was taken just sixteen months after the fire.

causes of these changes have been hard to disentangle, as they include everything from belated consequences of clear-cutting forests to megadrought to excessive pumping of groundwater, in addition to our overheating atmosphere, but researchers are developing better models that could help us understand our role. Because there's much more to this story, we will investigate the climate change part of it in greater detail in chapter 5.

Each year seems to host a competition to see what will come first: the hot, dry offshore winds that will fan catastrophic wildfires, or the first substantial rains that will extinguish the fire season. Even a light summer or early season rain can help quell the devastation of autumn fires. Billions of dollars and hundreds of lives are often in the balance during this drama of weather-pattern timing that often extends well into the autumn months. Some of the state's most disastrous conflagrations have been directly followed—within a few days—by soaking and

FIGURE 2-17. Recently burned and highly exposed drainage basins produce tremendous sediment yields when winter rains follow on the heels of autumn's devastating wildfires. During a winter storm, loose mud and debris flows were coughed out of this canyon in the Santa Monica Mountains and toward the beach around Leo Carrillo.

sometimes flooding rains that doused the fire season until the next year. Because the annual rainy season arrives on the heels of summer's drought and autumn's desiccating winds, drenching downpours may immediately fall on slopes just recently stripped bare by yesterday's fires. Billions of raindrops impacting the surface quickly merge into sheets of water and then torrents of overland flow that easily cut through steep, vulnerable slopes with loose materials and meager protective vegetation cover. Each year, nature sets the stage with a combination of these perfectly timed events and well-tuned actors to produce ideal flood and debris-flow environments within drainage basins that have some of the highest sediment yields in the world—all right here in California. We've seen these catastrophes play out too often in the media. Perhaps this explains why fire and flood have been described by some as California's only two seasons. "I've become an addicted follower of Doppler radar," quipped one woman living near recently burned landscapes transformed into mudflow country, who shared her predicament on the local news.

WIND BENEATH THEIR WINGS: USING METEOROLOGICAL SKILLS TO SURVIVE CHALLENGING TIMES

Droughts and occasional fires don't stop the annual butterfly migrations that flap and blow in with the wind. As autumn turns toward winter, mild coastal climates attract our renowned monarch butterflies (*Danaus plexippus*) to Central and Southern California. Each year, these delicate, colorful California (western) monarchs arrive from distant climates, most often from other western states where winters are cold. They camouflage well on Monterey pine and eucalyptus trees, sheltering from the sea breeze in dense clusters in the tree canopies. Multiple generations will travel up to 3,000 miles (4,800 km) while feeding on milkweed; single individuals have been known to migrate up to 650 miles (1,000 km). You might find them clustered in mild coastal shelters and canyons, such as at Pacific Grove and Morro Bay, from autumn through winter. They will take off on their journeys, most back north and east to those colder regions, as winter finally loses its grip on the continent. Fanciful fluttering begins each day after the butterflies sun themselves to loosen up in just the perfect range of temperatures. These may be the only insects that make such lengthy, round-trip migrations, and we're lucky to be a haven for their seasonal stopovers.

While many seed-collecting animals are scurrying to fill their winter stores, this can be a cruel season for their predators. The buffet might have been full of choices during the previous spring and early summer, but months of drought and heat can mean autumnal scarcity for hunting animals. You might spot eagles, one of California's fourteen species of hawks, and other raptors soaring in the thermal updrafts above most plant communities. They have the advantage of being able to fly away from drought and fire to where there is food and water. On one field trip, my students and I watched a red-tailed hawk (*Buteo jamaicensis*) soaring effortlessly until it suddenly dive-bombed to grab a snake dinner, clutching the writhing meal in its talons as it flew away. Among other, more common birds, turkey vultures (*Cathartes aura*) spend autumn feasting on animals that couldn't make it to the first life-giving rains of the season. Their bald heads are custom-made for hollowing out decaying meat and bones. In contrast to the majestic smooth soaring common to hawks, vultures fly on thermals with their wing tips turned toward the sky, tilting back and forth as if they were drunk. They might be sniffing out the gases given off by the carcass of drought's latest victim. I have seen them dancing in the wind in nearly every corner of the state.

California's most celebrated vultures and largest birds soared across the state and thrived by the thousands through a host of changing climates for thousands of years before they were nearly driven to extinction. By the 1980s, California condor (*Gymnogyps californianus*) populations had plummeted to fewer than thirty until they were all taken into captivity. They had been electrocuted on power lines, shot, and poisoned after eating leaded ammunition and drinking contaminated water that included radiator fluid. After the remaining few were nurtured and bred in captivity, condor numbers rebounded and are now being counted by the hundreds in the wild, with nearly two hundred in the Golden State. You might especially spot them gliding the thermals and searching for carcasses of the latest severe weather victims in the skies north of LA and well up into the Coast Ranges. Mature condors weigh about 20 pounds, with wing spans up to 9 feet (3 m), and along with their turkey vulture cousin are one of the few birds to rely exclusively on thermals and other updrafts to stay aloft for hours instead of using wing strokes to fly. It is a thrill to see this species redecorate our skies.

FIGURE 2-18. Our condors have the largest wing spans of any California bird and are the largest North American land birds. These soaring vultures that feast on decaying land and marine carrion (which helps to explain their bald heads) were miraculously saved from extinction by biologists.

California's iconic bald eagles (*Haliaeetus leucocephalus*) also nearly went extinct, plunging below thirty individuals in the state during the 1960s. They were victims of DDT and other pollutants, habitat destruction, and other human encroachments. More recent recovery efforts have been successful. By 2012, bald eagle numbers had soared to more than one thousand statewide, with more than two hundred bald eagles nesting across the state. Golden eagle (*Aquila chrysaetos*) populations have been more difficult to estimate. Most California bald eagles winter here to escape harsher climates, while most golden eagles remain in the Golden State year round. Each year they add to their enormous nests, sometimes until their homes crash to the ground under their own weight. Some nests have been measured at up to 6 feet in diameter, weighing more than a ton.

Many animals (including eagles) sense air pressure trends that can portend storms. Research has shown that eagles will soar above updrafts along topographic barriers, but that they can soar even higher (up to 10,000 feet [3,000 m]) within well-developed thermals on sunny afternoons or within a storm's ascending air currents. A few decades ago, my field studies students marveled at their first soaring eagle on a particularly strong thermal that had developed on a sunny afternoon just east of the Sierra Nevada. We watched the bird soar on the thermal's lift in an upward spiral directly over us, shrinking in our upraised binoculars, until it finally disappeared into the clear blue.

If you find yourself on the central coast where we started this wind-and-wings revelry section, stop at Morro Rock. Gaze beyond the scenic wisps of fog dancing around this igneous plug and you might spot a peregrine falcon (*Falco peregrinus*) nest perched on the cliff. If you are patient and lucky enough, you might observe one of these large, beautifully patterned falcons hunting other bird species in flight. They have been clocked at over 200 mph when they dive toward their prey, which makes them some of the fastest predators on Earth, which also makes them some of the best micrometeorologists on Earth. To achieve this speed, they have to ride thermals to soar two-thirds of a mile above the Earth's surface and then dive back down beak first toward a luckless pigeon or duck, flipping around in midair to strike it with their talons and declare lunch served. Every falcon carries Earth history lessons in tracking changing air pressure, wind, and other weather elements, dating back many generations. And here is another near extinction turned success story. It started when DDT ravaged the falcons' numbers until only five pairs remained in California in 1970. Today, as banned DDT continues to gradually dilute out of some of these marine environments, peregrine falcons have been thriving and are no longer threatened.

By following and studying the magic of Earth's flying species, we can better understand pressure patterns and air currents that inform us about our atmosphere. Pressure changes, updrafts, downdrafts, wind shear, eddies—just like clouds, dust, and smoke, the birds help make the air visible to us. And although they have the advantage of being able to fly away from California-style droughts and other climate stresses, they must rely on less fortunate species for their survival. Sounds a lot like us.

After earning his pilot's license, an old friend of mine from high school took me up for my first private plane ride during one of those post-storm afternoons. Our relatively smooth gliding was occasionally interrupted by sudden bumps that lifted the plane, until we emerged out of the updrafts that were usually marked by little puffy fair-weather cumulus clouds. It was easy to imagine how we had become like one of those high-flying raptors as we learned to sense and anticipate subtleties in the pressure changes and wind patterns. Pilots are expected to know some basic meteorology and then are briefed on current weather conditions before they take off. When you are bumped around by unsettling updrafts and downdrafts during your next flight, you should expect that your pilot understands what is causing the turbulence and how to avoid the worst of it. You may even have noticed how pilots sometimes bank around larger cumulus towers and storms that promise violent updrafts and downdrafts and potential white-knuckle danger that can crash a plane, especially during takeoffs and landings. In addition to interpreting the sky view ahead and monitoring updated weather information, they are often in contact with pilots who are in the same flight path. Technology has made us better birds. One day shortly after we enjoyed our flight, my old friend the pilot was forced to use bird-like precision to crash-land his sputtering small plane on an intersection in Fullerton, causing relatively minimal damage and no casualties.

CHANGING WEATHER PATTERNS SHAPE MOODS AND HOLIDAY TRADITIONS

By late November, we all expect changes in the weather. Northern Californians might experience the first turbulent storms or the return of enshrouding valley fog in colder, stagnant air; Southern Californians might still revel in those dry, warm, offshore weather patterns that can act as very different holiday mood setters. I have been lucky to experience all of these autumn-toward-winter weather patterns in every part of the state, but those clear 80° Thanksgiving Days stand out. In our amalgam of cultures here in California, we find ourselves gathering around tables laden with turkey, tamales, squash, sweet potatoes, pumpkin and blackberry pies, and other traditional family treats. But between the cooking and the eating, instead of gathering around a TV, we have time enough for a game of basketball or a bike ride or hike, thanks to our offshore breezes that so often serve up sunny afternoons. Such Thanksgiving Days that I have celebrated were often set against a backdrop of desert-dry winds funneling through the canyons and pouring out toward the ocean. People who live in California's coastal plains (especially in Southern California) are familiar with the alfresco moods set by these offshore winds. And when autumn's pressure gradients weaken and offshore breezes are gentle, nature coaxes people to go outside and take advantage of outdoor activities most other folks in harsher climates can only dream of during this time of year. Southern California's iconic beach boardwalks and bike paths become more navigable without the crushing summer crowds and street performers. The Golden Gate Bridge offers breathtaking views, and most coastal Northern California bike and walking paths are nearly as rewarding after the otherwise incessant fog has finally been swept away, just before winter rains set in.

These warm, dry offshore wind patterns also help explain scenes in the colorful cloth map that once hung in our bathroom, displaying the region's major landmarks and events that attract millions of people to Southern California. Out of about fifty illustrations, only two referred to indoor activities. The other forty-eight places of interest and suggested activities on the map ranged from the Hollywood Bowl and Rose Bowl to an assortment of athletic opportunities and venues, all outdoors. Southern Californians' obsession with outdoor recreation has been fueled by these periods of warm, sunny patterns that punctuate our autumns and winters.

But when autumn's land breezes grow much stronger, outdoor athletic escapades can be halted by more than blowing dust. Some of these autumn wind cycles have carried debilitating smoke plumes from upwind wildfires that have ignited and raged into December when the rains come late. The rust-colored smoke puffs suddenly erupt in the distance and can quickly grow into towering mushroom clouds. Santa Ana winds push the plumes toward the coast, over population centers, and eventually out over the sea. Once the fires start, local air quality will depend on your specific location in relation to the fire and the wind. As the plume drifts over, it first brings the indisputable (and unforgettable) smell of distant burning plant communities. The moon and sun dim to orange and then red behind the thickening smoke. Wild, out-of-control nature is visiting the neighborhood. During the day, everything might take on a red glow as in a foreboding apocalyptic scene from a Hollywood disaster movie. Just when you think it couldn't get creepier, ash and soot begin raining down. They fall in their quiet absurdity as fine and/or coarse particles, as if they were attempting to cruelly imitate the crystalline beauty and serenity of a snowfall. But the visible ash and cinder flakes that accumulate on cars and other surfaces may not be as dangerous to our health as the nearly invisible particulate matter known as ultrafine particles (UFPs). These tiny killers (with diameters less than 0.1 microns—millionths of a meter) can enter deep into the lungs with each breath and interfere with airway defenses. They carry toxic organic compounds and are so small, they can even be distributed from the lungs into the circulatory system. In these situations, athletic endeavors must come to a halt. Folks with preexisting respiratory problems are at greatest risk. Plans to celebrate health in the great outdoors are ruined. Adjacent regions only a few miles away may not be affected if the well-defined smoke plumes miss them and are carried out to sea. But the finer smoke often lingers out over the ocean until the onshore breezes return, spreading a film of sickening smoky haze over everyone and everything. As days pass, thin clouds of smoke can even mix with urban smog, which contains its own unhealthy particulate matter. I've watched these apocalyptic dramas play out so often, they have become anticipated mood changers each year. Such smoke attacks begin resembling uninvited, too-familiar guests who show up unannounced and stick around into the holiday season, wearing out a welcome they never had.

Mountain resorts and surrounding high country will also bask in brilliant November sunlight when these dry spells appear without wildfire smoke. But, at such high elevations, they don't benefit from the compressional heating. So, just as early freezes sweep through the northern highlands first, the cold eventually spreads into Southern California resorts from Wrightwood to Big Bear (nearby Holcomb Valley is often Southern California's coldest), and from Idyllwild to Julian. Mountain people can only look down on the warming breezes as high-elevation autumn temperatures continue dropping toward winter's hard freeze, and visiting hikers will need to take precautions.

Back in the crowded coastal flatlands, as November progresses toward December, chamber-of-commerce weather patterns finally begin to turn. The transition might begin with a gentle Pacific breeze, higher humidity, and high clouds streaming over from the west, all announcing major changes on the horizon. You might feel and smell the moisture as air pressures fall and Northern California begins to benefit from the first incoming drought busters. For thousands of years, the first people of California, who lived in close relationship with the land, anticipated and reacted to this annual transition from dry to wet. Today, we Californians try to analyze and predict these seasonal weather pattern changes that will finally begin to impact school and work and our various activities. We sow our native wildflower seeds so they will be blooming by spring. Our weather attention and vocabulary turn to cyclones, warm fronts, cold fronts, and shifting snow levels. (In this book, we use *snow levels* to mean the elevation above which snow will fall and below which rain will fall.) We can anticipate the soaking rains that will saturate our soils and water our ecosystems. We can hope for the beneficial runoff that will fill our reservoirs, revive our fisheries, and irrigate our farms. California's favorite son Don Henley sings us into autumn, seeing no one on the beach and feeling it in the air . . . the summer's out of reach. Like the Beach Boys, we may miss our warm, sandy beach days, but Californians who have been keeping a close eye on reservoir levels also enthusiastically welcome the rallying cry "storm watch" that sends us into winter.

FIGURE 2-19. Ominous unstable mammatus clouds above the LA Basin are signs of strong updrafts and downdrafts in the upper atmosphere that may be ushered in with winter's storms. Although mammatus clouds may be associated with severe thunderstorms, the clouds shown in this photo didn't have enough upper- or lower-level moisture to work with by the time they drifted over.

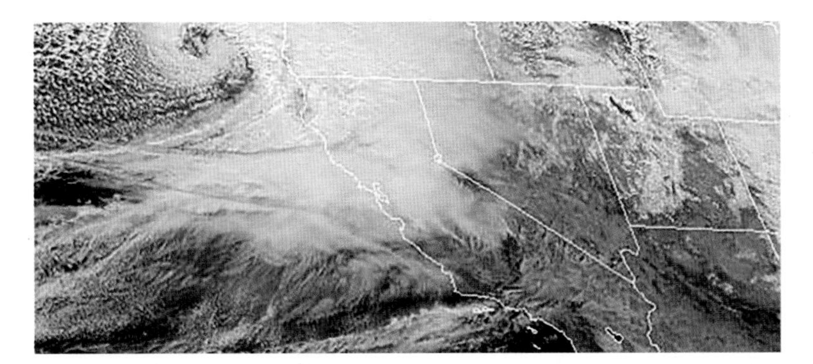

FIGURE 2-20. In this photo, we see a powerful cyclone bearing down on the West Coast near the Oregon border, signaling a sudden and violent transition from summer to winter weather patterns on November 26, 2019. As the storm pushed southeast, it quickly strengthened into a bomb cyclone. Crescent City reported the lowest pressure in California history as the wet and windy system circulated a front that eventually spread heavy precipitation across the state. *Source:* NOAA/National Weather Service.

FIGURE 2-21. This upper-level 500 mb map shows an exceptionally large and powerful low-pressure trough that is pinching off from upper-level winds and pressure patterns (Nov. 21, 2019). It has drifted over Southern California, perhaps as an extreme example of a cutoff low that has dropped farther south and intensified far beyond what might be expected from average inside sliders, which more commonly remain just east of California. *Source:* NOAA National Centers for Environmental Prediction, Weather Prediction Center.

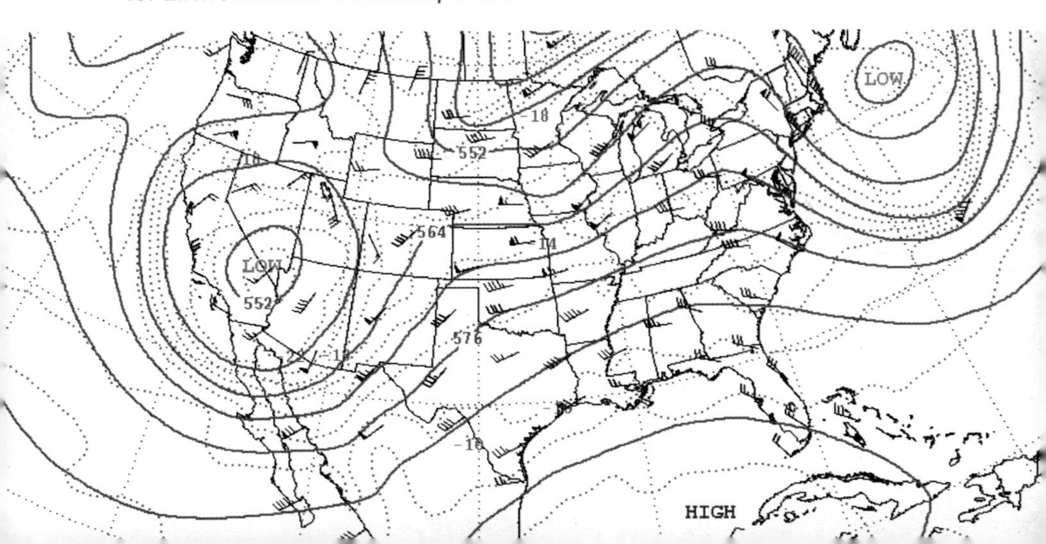

FIGURE 2-22. As autumn turns toward winter, Pacific storms should begin sweeping over the San Jacinto Mountains, delivering replenishing rain and snow to quench the drought and end the fire season. Here at Lake Fulmor, an early winter storm brings upslope fog and abundant precipitation that may later turn to snow. Colder winters may freeze the lake surface.

FIGURE 2-23. A blanket of tule fog is slowly burning off after settling in the Central Valley on December 4, 2021. Such radiation fog portends the long, cold nights that frequently follow winter's transient storms. The misty cotton merges in the Bay Area with a thick marine layer (more reminiscent of spring and summer) that covers most of the California coast. Note how serpentine ribbons of fog have also settled in the Klamath and Trinity River Valleys near the Oregon border. *Source:* NOAA/National Weather Service.

CLOUDS
HARBINGERS OF SEASONAL CHANGE

Clouds play major roles throughout this book, so we should pause here to present this helpful cloud interlude. It is a very brief but colorfully informative rest stop along our seasonal tours across the state.

Clouds do more than add beauty and magic to your sky dome: they offer lessons and tell you stories about how the atmosphere and seasons are changing. I have often reminded my students that clouds are signposts in the sky, demonstrating the forces and physical laws that are at work in otherwise mostly invisible air. If the high- and low-pressure systems drifting over you came with giant Hs and Ls in the sky and you could see invisible air moving, it would be easier to understand the changing weather and seasons. Instead, you are left to interpret and analyze the shapes and movements of smoke, fog, and clouds when you don't have sophisticated weather instruments to measure all the elements. Good thing you have those transitory clouds. Unlike the constellations, specific cloud formations don't appear and reappear at the same exact places in the sky on the same dates every year. But if you look more closely, you might be able to pick out cloud characteristics and behaviors common to weather patterns of particular seasons. Can you see the signs of changing weather and seasons in the clouds? Can they tell you whether we are going to have an early or late winter?

Have you ever noticed how clouds often appear more overcast on the horizon? This may be because you are surrounded by mountains that encourage clouds to build in the rising, expanding, and cooling air currents that must ascend over the summits. But these scenes could also be illusions cast by your perspective as you look into the sky: it is simply easier to peer through the openings and breaks between scattered clouds directly over your head. As you look farther away toward the horizon, you must view the clouds at angles and more toward their sides; distant openings and breaks are obscured between and behind clouds that appear to be much closer together than they are. These deceptive angles can be frustrating for someone who prefers cloudier skies and potential stormy weather that often appears so close, but never arrives. It may seem that you have only scattered clouds and partly cloudy skies above you while your neighbors are under constant broken or overcast ceilings—lasting illusions of winter over them, summer over you.

Or could these cloud ceiling deceptions represent a metaphor that offers insights into your life attitude and philosophy? Is it partly cloudy or partly sunny? Is the sky half full of clouds or half clear? Instead of bemoaning the shadows clouds cast, you can find the beauty in their flow and shape; nature has her own plans and laws, and you will have a lot more fun and find a lot more satisfaction by going with the flow, objectively preparing for likely weather events and seasonal change instead of hoping the forecast will change. Enjoy the sky-show surprises. The clouds are talking to us. They are showing us the way to the next season.

When I was a kid, I would lie on the grass and watch clouds run into one another. Someone told me that thunder was caused by clouds colliding. As I grew a little older, I realized—on a day when vertical wind shear blew lower clouds one way and higher clouds in another direction—that the clouds I was watching were at different levels. They only appeared to be colliding when I viewed them up through the different layers. I also later

FIGURE 2-24. Is there a cumulus congestus pileus in your neighborhood? This boiling cloud is developing as moist air rises, expands, and cools to its dew point so that latent heat can then fuel stronger updrafts. Notice the silky cap (pileus) cloud that has formed on top as upper-level currents are forced up and over the rising air parcels. The rainy season has arrived in coastal California.

learned the truth about thunder: as discussed in chapter 1, it is a shock wave produced when superhot lightning discharges in billowing cumulonimbus clouds (thunderheads), which causes the sudden explosive expansion of air molecules.

MEASURING AND TRACKING CLOUDS FROM DIFFERENT PERSPECTIVES

Clouds at different levels can tell you how much moisture exists at different altitudes and whether that moisture is reaching its dew point in a stable or unstable atmosphere. Flat (stratus) clouds suggest that a layer is saturated but also relatively stable. Vertically developing puffy (cumulus) clouds suggest there is an unstable, turbulent layer. It is common to find one layer stable and another unstable. Examples include low stratus clouds drifting onshore in the stable marine layer, below middle-level castles of altocumulus (altocumulus castellanus) clouds or higher cirrocumulus clouds that might display upper layers of moisture and instability. Or low-level cumulus clouds (often billowing in moist thermals rising from the surface on a warm afternoon) could begin ballooning up toward flat middle-level altostratus or higher cirrostratus clouds that signal layers of stable air aloft.

When you look into a sky with broken clouds at different levels, there could be several thousands of feet of clear, dry atmosphere between the layers. For instance, you might be able to view high ice-crystal cirrus clouds through open breaks in low stratus clouds. An entire clear column of atmosphere with unlimited visibility may exist between the low cloud layer at about 2,000 feet (several hundred meters) all the way up to the cirrus clouds at 25,000 feet (7,600 m) or higher. A better way to make these observations is during your next plane flight through the cloud layers. You might fly though the fuzzy flat stratus clouds that often signal innocuous stability. They contrast with the well-defined puffy, unstable cumulus clouds with bright cauliflower outlines. Such distinct edges mark the boundaries between rising air within the cloud and sinking clear air adjacent to it. Your plane is likely to be bouncing around in this turbulence. Don't be too annoyed if you end up in a seat close to me on a flight. I will be the passenger who keeps the shade up on my window so that I can look out at the changing cloud patterns and the terra firma exhibitions below.

When on the ground, I have noticed that cloud formations appear more defined, and visible layers and other optical phenomena seem more distinct, when I'm looking at them upside down. I discovered this quirky and rather startling perspective while hanging from my legs on pull-up bars, stretching my back after some grueling basketball games. Make sure you know what you're doing before trying this, as your head will be the first to hit the ground if you fall. Could such falls help explain my obsession with the weather?

You can try estimating the distance of clouds by using simple triangulation. This is a crude exploitation of more precise and sophisticated measurements made by advanced remote sensors and even today's lower-cost, high-resolution digital cameras. Look out and estimate the angle of each cloud above your horizon. Clouds closer to your horizon line are farther away, and those closer to your zenith point (directly above) are nearly on top of you. Look carefully to estimate the height of the cloud and draw an imaginary line straight from the cloud down to the ground directly below it. Even if you are looking near the horizon, a cloud very low to the ground may be only a mile or so away. But if you are looking near the horizon at a cirrus cloud or the anvil top of a cumulonimbus cloud that may be above 25,000 feet (7,600 m), such a high cloud could be as much as 100 miles (161 km) distant. Who might be looking straight overhead to see your distant high cirrus cloud, and where might they live?

In winds of about 10 mph, the lowest clouds might drift from the horizon and over you in a few minutes. But in 10 mph winds at 25,000 feet (considered a low velocity for upper-level winds), it might take 10 hours for those distant high cirrus clouds to finally drift over your head from the horizon. At the same wind velocities, low clouds will appear to race over your head, since they are closer; but higher clouds moving at the same speed might seem to be barely moving at all, since they are so far away. This is similar to when you stand next to a road and observe how the cars whiz by at such high speeds; but when you view vehicles at the same speeds on a distant highway, they seem to be just creeping along. It is likely that you will notice those high clouds moving a lot faster as winter approaches, as the high-velocity jet stream migrates south, closer to California.

NIGHT MOVES REVEAL NOCTURNAL AND SEASONAL MOODS

You can use a few tricks to track clouds at all levels at night, but why bother? Because nocturnal weather performs through half of the year and half of your life. Because exciting weather doesn't go to sleep at night, and you can remain curious about the changes that continue up there in the darkness. In remote locations with little light pollution, the clouds will appear as dark blobs obscuring your view of background stars and planets. Or they may be slightly illuminated by the moon, allowing you to make out their shapes and structures. Stand still to track their silhouettes as they drift like dark ghosts with the upper winds. By contrast, if you live in or near a city, the cloud bottoms will be brightly illuminated by the lights, while their backgrounds

FIGURE 2-25. Grazing cattle look up to high-altitude contrails (condensation trails) in Bridgeport Valley just east of Sierra Nevada's glaciated terrain (distant far right). We're only a few days away from the first snowfall on this autumn day.

become relatively darker. Trace their outlines and follow these brighter ghostly patches as they flow with the wind above your well-lit neighborhood. You are now observing the weather twenty-four hours each day without any instruments. There's no charge for these nocturnal shows. Funny how the night moves.

We have made amazing technological advances since I was a kid honing my observation skills by comparing my cloud height estimates with the experts on aviation and the weather radio recordings. You can imagine why this knowledge can be so useful besides being just for fun. Today, cloud base heights (referred to as ceilings when broken or overcast) are measured using laser-beam ceilometers. Laser-beam pulses are shot up to the clouds, where they reflect back to surface sensors. Cloud heights are measured by the time it takes these electromagnetic radiation pulses to reflect from cloud bases to ceilometers back on the ground. And they are working day and night! Such cloud-base snapshots are a common part of updated weather conditions reported from most major weather stations, especially at airports; you will find them conveniently posted and updated on National Weather Service websites.

BLAZING A JET TRAIL

We can thank aircraft for making clouds that enable us on the ground to estimate wind velocities and weather patterns way up high and also help us understand how clouds can form at very cold, high altitudes. When you see jet trails—or condensation trails (contrails)—forming, they are announcing that the upper altitudes, where jets are flying, are near their dew points. You might notice them during any season, but once they form, they are likely to drift faster with higher-velocity winds as nature's winter jet streams sweep farther south. Jet planes emit particulates and moisture in their exhaust into air at around 30,000 feet (9,140 m) altitude. Way up there, where it may be colder than −50°F (−46°C), vapor will almost immediately freeze around the jet exhaust to form ice-crystal cirrus-cloud streaks. When jet trails are thicker and last longer, there are often cirrus clouds forming near them in the saturated air at these high altitudes. When drier upper layers are not near their dew points, the jet trails will quickly sublimate (turn from ice directly into vapor) and disappear into the clear air. You can even track specific airline flights that might be making trails; try researching their flight patterns. Research has shown that these slender clouds can combine to block and reflect enough shortwave sunlight back to space to suppress afternoon surface temperatures downward by a degree or more. When conditions are favorable, look for the linear shadows they can cast through hazy skies. But they can also absorb longwave radiation from Earth's surface at night, only to reradiate it back toward the surface, keeping overnight low temperatures just a bit warmer. The net result might be slight global warming in our atmosphere. At these high altitudes, the suspended jet exhaust pollution will likely be carried hundreds or thousands of miles in strong upper-level winds until it gets diluted and dispersed into global circulation patterns.

BEAUTY OR BEAST?

We all have tried to assess cloud threats by analyzing their different shapes and shades. Why are taller, denser clouds with more moisture a lot darker on the bottom? This is because they are very efficient reflectors *and* absorbers of radiation in the visible part of the spectrum. Denser cumulus and cumulonimbus clouds may appear very bright and silvery on their sides and tops. So much solar radiation is reflected by or absorbed within these tallest clouds that it can become almost as dark as night at Earth's surface beneath the cloud. Anyone who looks up at the bottom of one of these angry-looking behemoths knows to seek shelter from the coming storm. However, tall stacks of thick stratus cloud layers can also block enough solar radiation to appear quite dark on their bottoms. These harmless but menacing-looking fuzzy sky blankets of stratus clouds may only produce tiny raindrops or fine drizzle or no precipitation at all.

The first raindrops from an overcast sky can help us identify the types of clouds and their intentions. Mist and microscopic drizzle drifting around suggest stratus clouds that pose little danger. Heavy, gargantuan drops are likely to be falling out of threatening, turbulent thunderheads. Powerful updrafts lift and buoy ice crystals high within such cumulonimbus clouds. Suspended ice rocks continue to attract abundant moisture, which will freeze on to them within the updrafts until they can grow into giant hailstones. The hail eventually grows too heavy and falls to Earth or is hurled to the surface after getting caught in violent downdrafts. In California, these hailstones often melt as they fall toward lower elevations, but they will make big impacts when they kerplop as giant raindrops. When larger hail falls out of particularly cold air in high-velocity downbursts, it may not have a chance to completely melt. This usually results in pea-sized hail in the Golden State, as opposed to the hail from out-of-control monster thunderstorms that can build in the Plains and Midwest starting in spring. Such storms are very rare in California since we lack the dramatically contrasting air masses and other weather patterns required to energize comparable sky wars. Cumulonimbus clouds, with their thunderstorms and hailstorms, appear over coastal California more often during winter and early spring months because the state's dominating stable high pressure usually squelches such turbulence during summer. An autumn thunderstorm on the coast can be the first signal that winter's instability has finally penetrated though summer's resilient high-pressure ridges.

In November 2003, an upper-level low-pressure system produced isolated thunderstorms around the LA Basin. One of the storms boiled up and exploded during the afternoon, stalling over South LA communities that included Compton and Watts. In addition to frequent lighting strikes, damaging downdrafts, and more than 5 inches of flooding rain in less than 3 hours, it produced a historic hailstorm. More than a foot of hail fell on a few neighborhoods, accumulating in high piles that halted traffic and interrupted the daily routines of astonished residents. Meteorologists considered this an outlier event that occurs at an average frequency of more than 500 years in that region. It was an exceptionally ominous cumulonimbus cloud that lives in our history books, when fall turned into winter in one day.

The familiar phrase *head in the clouds* has appeared in countless poems and songs that continue to be popular to this day. Now that your head is up there, you can use your observations to follow the changing weather patterns that lead from autumn into winter. The long, hot days are far behind us, and it's time to welcome the short days of life-giving rain and snow.

FIGURE 2-26. Nimbostratus clouds dump substantial snow on higher slopes around Lake Tahoe, but only flurries make it down to the lake surface on this day. Surrounding mountains and the lake both serve to modify air masses and weather patterns, creating distinct microclimates.

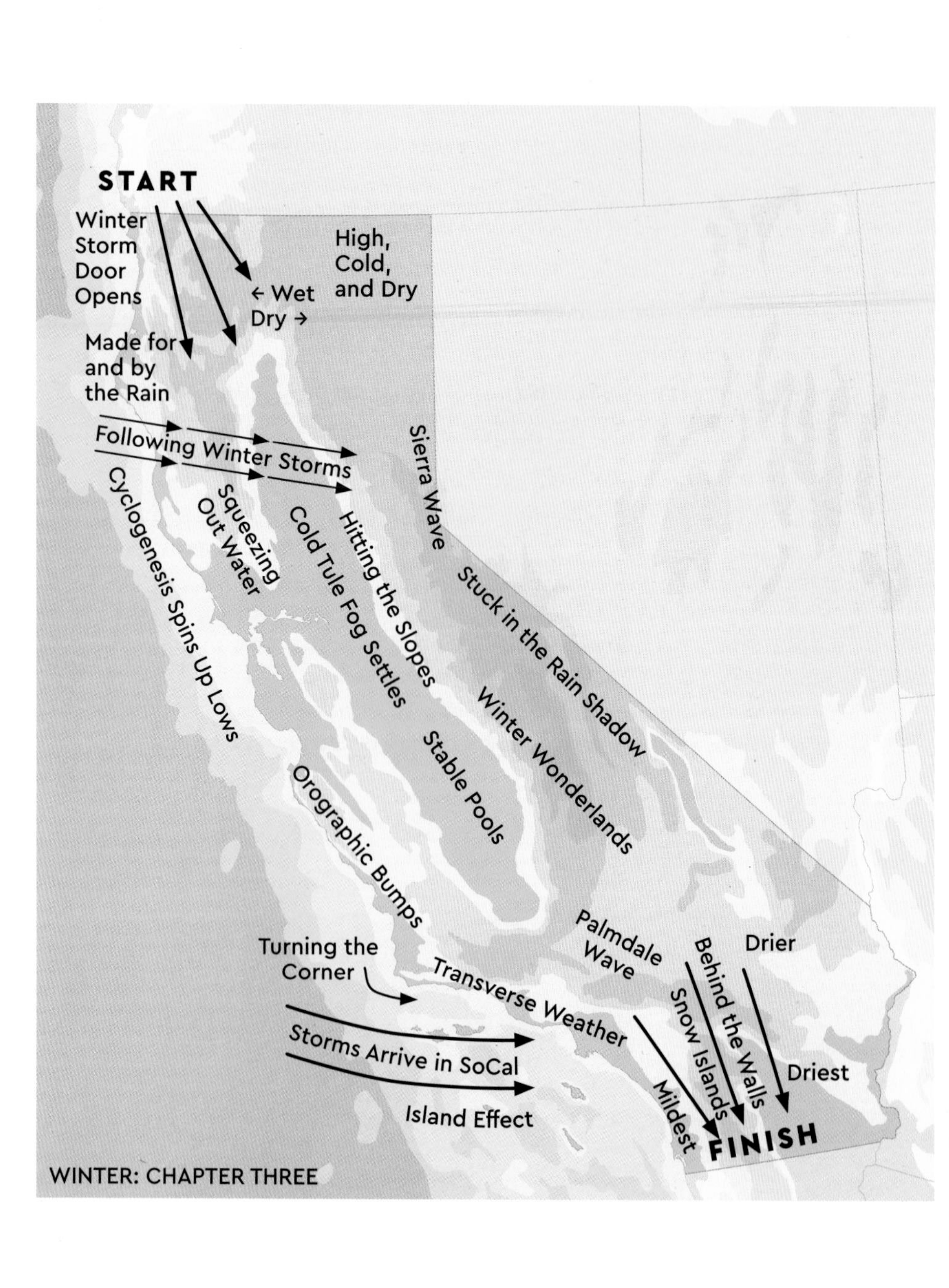
START
Winter Storm Door Opens
← Wet
Dry →
High, Cold, and Dry
Made for and by the Rain
Following Winter Storms
Sierra Wave
Cyclogenesis Spins Up Lows
Squeezing Out Water
Cold Tule Fog Settles
Hitting the Slopes
Stuck in the Rain Shadow
Winter Wonderlands
Stable Pools
Orographic Bumps
Palmdale Wave
Drier
Behind the Walls
Snow Islands
Turning the Corner
Transverse Weather
Storms Arrive in SoCal
Driest
Mildest
Island Effect
FINISH
WINTER: CHAPTER THREE

CHAPTER THREE

WINTER'S NOURISHING TURBULENCE

WINTER'S STORM DOOR OPENS: GOODBYE TO ENDLESS SUMMER

Wintertime is calling. The constellations Orion and Taurus, Canis Major and Minor, and Gemini return to greet those of us who venture out in our layers and warm jackets to admire them during the long, dark nights. We start our winter-season tour of weather patterns on our wettest northwest coast, where North Pacific storms usually arrive first and are most frequent. We then work our way southeast across the state, toward the drier winter climates along the southern coast, until we finally land in the driest rain shadows of our southeastern deserts.

The high sun has gradually retreated south, casting longer shadows, and the days grow even shorter as autumn yields to winter. The great summer protector, that North Pacific Subtropical High, unhurriedly but sometimes spasmodically weakens and also drifts south. At upper levels, the jet stream

FIGURE 3-1. Rainbows become more common on the coastal sides of our great mountain barriers during the unstable winter rainy season.

follows, dropping out of the Arctic; this opens the door to the first big storms, often associated with that expanding low pressure and jet stream pushing south out of the Gulf of Alaska and off the North Pacific Ocean (as discussed in chapter 2). These cyclones are defined by their counterclockwise spin and stormy weather, and we will soon see why scientists describe them as mid-latitude wave cyclones. They might arrive in dramatic fashion, sometimes as "bomb cyclones" (powerful low-pressure systems that suddenly strengthen), bringing gale-force winds, the first heavy rains along the coast, and the first major snows to the northern Cascades. Since they are so often caught directly in the bull's-eye of winter's storm track and jet stream, residents in Washington and Oregon recognize how these atmospheric upheavals end their comfortable fair-weather days, which inevitably yield to the relentless rain and snow that help define Pacific Northwest winters. And these Pacific cyclones will occasionally sweep into California through the next few months. Everyone (especially rural folks

FIGURE 3-2. An unusual flash flood spills out across the beach as thunderstorms move onshore. Such frequent lightning (notice the cloud-to-sea surface bolt) and drenching cloudbursts are rare on the California coast. The destabilized atmosphere warns that summer's monotonous fair weather is yielding to winter's occasional turbulence.

and those working and playing outdoors) must relearn how to cope with these early storms as they encroach south and erase whatever remains of summer, forcing a different way of living and dressing that requires layers of protection and careful planning before venturing out.

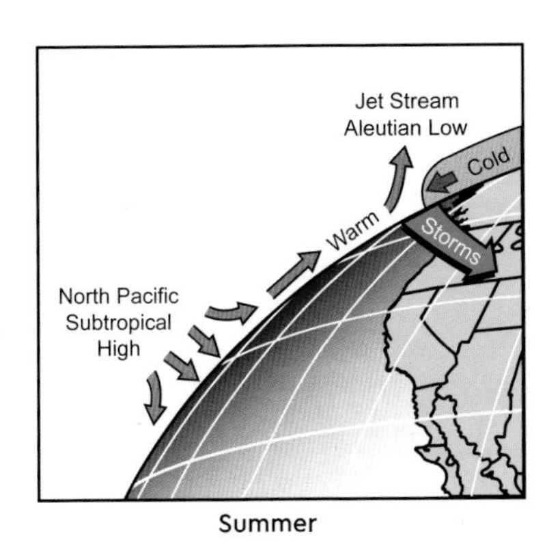

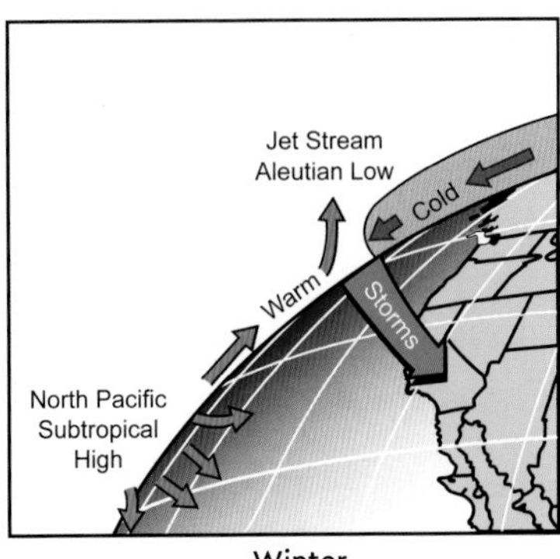

FIGURE 3-3. The jet stream and storm track usually shift south toward California as winter arrives.

MADE FOR AND BY THE RAIN

As winter storm systems penetrate farther south, they eventually drag into Northern California's relatively sparsely populated natural landscapes. Treacherous seas begin challenging fishermen, who might be forced to hunker down in harbors, such as at heavily fortified Crescent City. Natural environments are waiting for the incessant seasonal storms that quickly rehydrate the soil until runoff again engorges streams and rivers. Northwest California geography is made for and by the rain, from the misty shoreline where moisture condenses into haze around salt spray and the forest meets the beach, up to the massive coastal rock foundations of spectacular mountain ranges blanketed by lush plant communities. Below the pervasive drizzle caught by steep mountains grow thick forests with coast redwood (*Sequoia sempervirens*), the tallest trees in the world. Their formerly wider range retreated to these soggy coastal slope strips as warmer and drier conditions ended the Ice Age. Today, they thrive where you will find California's shortest summer droughts, heaviest rains, and most consistent fog, but without long periods of bitter winter cold. The perfect combination of temperature, moisture, and sunlight cycles (and other factors such as soils) enable redwoods to successfully compete with the diversity of species that also flourish in the cool moisture of these low-elevation forests. Researchers have postulated that the redwood's natural range could extend farther into Oregon as climates warm.

The summer fog helped to keep evapotranspiration rates low and added fog drip to dampen surfaces; a continual series of winter storms will now saturate the soil and feed the waterways. Here is where the greenest north coast coniferous forest communities (commonly found along the wet Pacific Northwest into Canada) merge with endemic coast redwood forests. The cool, musty air swirls and settles on these shaded slopes beneath the tallest redwood trees. And it's not just the trees that are exceptional. Here you will find some of the greatest total mass of living organisms on land (what biologists call terrestrial biomass) of any California plant community. Such diverse biomass includes slimy yellow banana slugs that produce mucus to help them glide through dense tapestries of fern species on and above the forest floor, where direct sunlight may never penetrate through the thick canopy. As winter arrives, this wet place gets wetter; the dull gray of summer's perpetual fog is replaced by erratic winds and inundating rainstorms that seem to soak through the skin. In far northern California, the cold and damp are nurtured by long nights and short days when the noon sun angle barely and only briefly creeps to just 24.5 degrees above the southern horizon, casting long shadows and lazy hibernal moods even at midday near the winter solstice.

When Pacific storms roll through and become more frequent as winter progresses, their moist, unstable air parcels are forced to rise up these coastal mountain slopes. As the air is pushed higher, it is forced to expand and cool to its dew point, resulting in heavy rain along windward slopes and heavy snow as moist air glides over mountain barriers. We're talking about an average of more than 100 inches (254 cm) of rain each year on the wetter slopes facing the Pacific. That creates ecosystems thriving in a world of water that can only be imagined farther south or in California's rain shadows. These dense forests are where the first Bigfoot legends arose. Also known as the Emerald Triangle, these forests have hidden and sheltered some of the most profitable and infamous marijuana farms for decades. On these shady slopes, growers have illegally rerouted water courses, cut trees, poisoned the land and wildlife with their dangerous chemicals, and even murdered bandits and innocents to protect their clandestine profits. (Cannabis users who don't grow pot in their own containers are urged to purchase from legitimate retailers and pay the sales taxes.) The saturated slopes rise up and come together to shape the mighty, rugged Klamath Mountains and Northern Coast Ranges. Powerful streams and rivers charged by runoff from winter's persistent downpours have cut deep, narrow chasms into these exceptionally steep, formidable mountain terrains.

Even the most annoying extroverts are encouraged to draw inward, settling in their nearest warm and dry cocoons. Stay-at-home hobbies and chores rule, the kettle is always on the stove, and the blues are on the stereo all season long. Unlike the summer's gray that hugged only the immediate coastal fog belt, winter's cold and damp can spread deep through interior regions, up into the mountain ranges and down into inland valleys. For folks farther from the coast, discussions about blazing summer sun and heat have turned to guessing snow levels for incoming storms. It is not unusual to see some form of rain or snow showers appearing every day in a Northern California seven-day winter forecast. But unlike summer's coastal weather tedium, major breaks may appear in the clouds. When ridges of high pressure drift over from the Pacific behind transient low-pressure systems, brilliant sun and crystal-blue skies are cherished but fleeting surprises. Because there's so much instability and changeable weather as high pressure nips at the heels of the storms, these bright periods are frequently peppered with strong pressure gradients driving brisk winds that can chill to the bone. Blustery days become the norm. Wet storms usually dominate the forecast farther north, while the sunny breaks become more frequent as you travel farther south.

About a lifetime ago, I took about ten days off work before starting graduate school so that I could nurse my unreliable car up along the spectacular redwood coast. While camping near the tall trees at Prairie Creek Redwoods, I heard the patter of raindrops splattering on the tent all through the night and listened to the occasional rumble of distant thunder; these sounds signaled the beginning of another long rainy season. Much more recently, during another of my many adventures in California's northwest, we bedded down in the dead of winter at little Gold Bluffs Beach Campground, which can flood during winter's storms. It is usually so wet there that the forest grows downhill to the beach, and trees anchor on isolated rocky sea stacks poking out of the ocean. Nearby Fern Canyon is so lush that it was used as backdrop for the Jurassic Park movie. But we arrived when California was in the grips of one of its epic droughts. This was one of those rare winter periods when exceptionally resilient high pressure dominated, delivering perfect hiking weather and an awesome unobstructed sky dome displaying thousands of stars during long, cold nights. Such a long dry spell, in the middle of the rainy season within this landscape of rain, offered us exceptional comfort and convenience, but left us with an eerie feeling that something was wrong.

HOW OCEAN CYCLES CHANGE WEATHER AND CLIMATE

Wild swings in California weather patterns that include the state's wettest and driest years are often attributed to changes in ocean temperatures. Researchers blame both natural oscillations and the increased warming of the land, atmosphere, and oceans caused by air pollution. This book and project are a celebration of natural seasonal cycles and the connections between the ocean and the land as they interact with and impact changing weather elements such as temperature and moisture. In the final chapter, I examine in more detail the impact of human-caused climate change on our weather, which is already destabilizing previously more predictable patterns.

To help you understand why there was a clear, bright sky in Humboldt County at that dry campsite as December turned to January, I'll start the discussion with the longer-term cycles that sweep across the oceans and impact annual weather patterns. Scientists have been measuring and analyzing these changes so that they have a better grasp of the ways that complicated relationships between the atmosphere and oceans are affecting everything on land, including us humans. In California, two of the most researched atmosphere–ocean interaction change agents (and heralded acronyms) make significant impacts on large-scale weather patterns: ENSO and PDO. You may not recognize the acronyms, but you're probably already familiar with their impacts.

ENSO (El Niño–Southern Oscillation) cycles have orchestrated some of California's most exceptional weather patterns and catastrophes. During average years, the low-pressure area near the equator (known as the Equatorial Low or Intertropical Convergence Zone) draws trade winds in to fill that void. In the Southern Hemisphere, winds turn out of that subtropical high-pressure zone and toward the equator as southeast trade winds. In the Northern Hemisphere, winds pour out of the North Pacific High and down the California coast, eventually turning back out to sea and toward the equator to become the northeast trade winds. These winds normally push warming ocean-surface water west across the Pacific and pile it up into the tropical West Pacific Ocean. In contrast to the cool, stable weather along the California and northern Baja coasts, warm, humid, unstable weather with heavy rain and tropical cyclones are common in the western Pacific (figure 3-4).

Every few years, the Equatorial Low weakens, pressures rise in the West Pacific, and the trade winds also fade. The cold Humboldt Current along the coast of South America and our California Current may slow. During these El Niño events, pools of warmer water begin backing toward Central America and spreading into the eastern Pacific near the equator, also causing sea levels to rise there. As water temperatures and pressure patterns reverse from the average, unstable stormy weather also sloshes into the eastern Pacific. Heavy rains and flooding often result along the South and Central American coasts and may sometimes even extend all the way north toward Southern California. Because South American fishermen noticed how these changes that could destroy their fishing season seemed to occur around Christmastime, they labeled it with the term *El Niño*, "the boy child." It seems to recur in periods of 2–7 years.

Although those heavy rains far to the south may not directly impact California, our state often experiences the indirect impacts of an El Niño. The huge contrast between that super-warm water and tropical air to the south and the colder water and air in the far North Pacific tends to draw the jet stream farther south and over California. When the annual winter storms form in the North Pacific, they are often guided toward California by that enhanced jet stream. As the storms approach the coast, they can tap the extra energy and moisture from unusually warm eastern Pacific air masses, pulling them into their cyclonic (counterclockwise-spinning) winds. The invigorated storms may gather and transport this tropical moisture ahead of them and their frontal systems as they sweep heavy precipitation across especially Southern California. Think atmospheric rivers (long bands of concentrated moisture streaming across the ocean from the tropics) on steroids. Among fifteen El Niño events within sixty years, eastern Pacific sea-surface water temperatures peaked at an average of about 3°F (1.5°C) higher than normal. The two strongest, much warmer, and most classic El Niño events (before the 2015–16 El Niño) occurred in 1982–83 and 1997–98, and each resulted in record precipitation

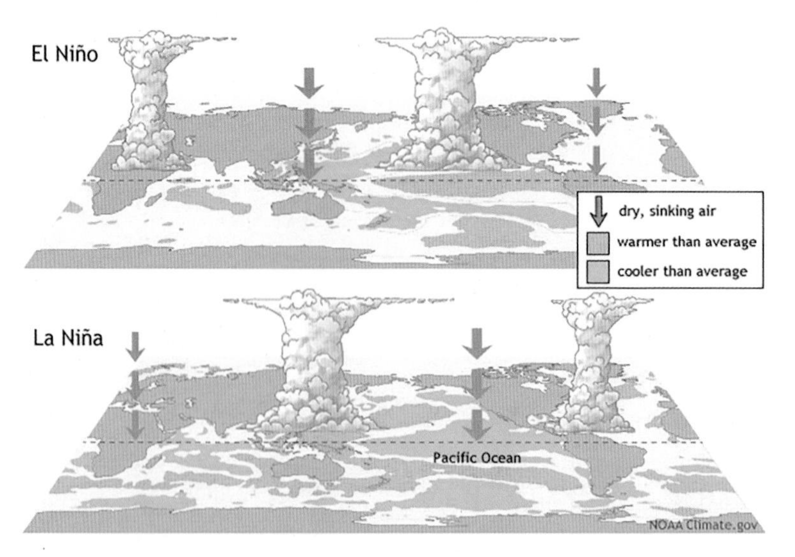

FIGURE 3-4. During most El Niño years, warm ocean temperatures pool toward the eastern Pacific, dragging low pressure and unstable, stormy weather toward the West Coast (from Southern California and south). During most La Niña years, stronger cold currents in the eastern Pacific (such as the California Current) push warm waters away and into the western Pacific. Strong La Niñas often leave West Coast regions (from Southern California and south) with stable drought while wet, unstable convection builds in the far western Pacific. *Source:* NOAA Climate.gov.

and flooding in California. As of this writing (2023), a strong El Niño event appears to be forming, so you will know the results if you're reading this years later.

The 1982–83 El Niño caught Californians by surprise, causing unexpected devastation from flooding, high waves, and storm surges. Scientists learned from such events and accurately predicted the 1997–98 El Niño by measuring early warning signs over the Pacific Ocean. Advance preparation, such as clearing flood channels, saved billions of dollars in damage and scores of lives. Many communities from Northern California to Orange County measured their wettest rainy season in recorded history, including all-time records in San Francisco and Santa Barbara.

However, just as every storm and weather event is unique, not every El Niño results in heavy rains in California. Even when the expected rains arrive, the focus of the precipitation will depend on the specific location of atmospheric rivers that drape themselves across parts of the state and deliver the water like narrow firehoses. Although the 2015–16 El Niño was powerful, the timing of storms and precipitation in California did not match the measured peaks in the Pacific, as it did in the 1982–83 and 1997–98 events. Scientists are studying other possible variables that might create lag times and various exceptions to expected ENSO patterns and behaviors, and their effects on California.

So some El Niño years are our super-wet years, when Californians both celebrate the rain and worry about its torrential impact. Approximately 60 percent of El Niño years are followed by La Niña years. In this reversal of fortune, the Equatorial Low pressure deepens, trade winds increase, and the California Current may become more powerful, carrying colder water down the coast and into the eastern Pacific. While the western

Pacific heats up and storms intensify there, our nearby eastern Pacific Ocean surface water becomes colder than average. This often forms cool, stable air masses over our coastal waters and can encourage the formation of those infamous storm-blocking resilient ridges that spell drought, especially for Southern California. During La Niña, atmospheric rivers may not be as powerful, and winter storm tracks may be nudged farther north. Consequently, these changing ocean current cycles change expected rainy-season patterns in ways that surprise campers and others venturing into the great outdoors during winter.

There are many exceptions to these rules and expectations in a weather world of averages, probabilities, and endless variables, and every weather pattern and event is at least somewhat different from the others. Nature can be quite a trickster. One glaring example is the winter season of 2023, which marked the third straight La Niña year; this phenomenon is so rare that it earned the title "triple-dip La Niña." Exceptional flooding rains and snowfall across the state peaked with record rainfall in Central California and was capped by some of the highest snowfall totals ever recorded in the Sierra Nevada. Catastrophic drought turned to inundating floods within weeks in this El Niño–style rainy season at the end of a historic La Niña. Human-caused climate change may also be intensifying both the dry and wet extremes of the ENSO cycles. As global temperatures climb, these cycles are likely to deviate further from the norm, making the weather much harder to predict as the patterns of the past change in unexpected ways.

Recent research indicates that ENSO may be the greatest single natural large-scale weather influencer on Earth. These oscillations have been blamed for weather anomalies around the world, such as a decrease in the number of hurricanes in the Atlantic Ocean during El Niño events. Numerous other far-reaching cycles, such as the Pacific Decadal Oscillation (PDO), are also gaining attention as scientists learn more.

PDO is another cycle that changes ocean surface temperatures across the Pacific Basin, rerouting the flow of the jet stream and storms. In contrast to ENSO, PDO patterns may last for a few to several years in overall cycles of 20–30 years, and the undulations of the PDO can weaken ENSO storms and droughts, or strengthen them, depending on the phase. As sea-surface temperature changes gradually sweep across the Pacific, a warm ocean water phase usually coincides with lower surface air pressure over the North Pacific, resulting in more rain in the American Southwest. When those temperature and climate patterns are reversed (in the "cool phase"), above-average surface air pressures in the North Pacific might leave us high and dry on land. PDO cycles help determine whether the California coast might be stuck under the wet sides of upper-level troughs or the dry sides of upper-level ridges.

When we consider the many other seemingly infinite overlays of long- and short-term and large- and small-scale cycles in our atmosphere and oceans, we can appreciate the complexities that researchers are struggling to unravel. Just one more example is the shorter-term Madden-Julian Oscillation (MJO), which sweeps along on a weekly time scale. MJO's connections to atmospheric rivers from Asia across the Pacific and periodic flooding along our West Coast were on display again during the winter and spring of 2023. These cycles will continue even as the climate and oceans continue warming, but the impact of all that extra heat in the atmosphere is destabilizing them in ways scientists are just coming to fully grasp. In chapter 5, we explore how climate change affects current weather and could shape the future weather of California. The howling winds, ferocious waves, and flooded towns of winter 2023's series of atmospheric rivers brought Californians a sobering premonition of what some winters might look like as more heat energy accumulates and cycles through the ocean and air around us.

FIGURE 3-5. Northwest California's wet western slopes are covered with coast redwood and other forest species that thrive in heavy winter rains and cool summer fog. These are very moist microclimates and plant communities.

MOVING TO THE DRY SIDE

Because much of the moisture is wrung out of winter's Pacific cyclones as they pass over the Klamaths, the slopes and plant communities change noticeably from wet to relatively dry as we travel toward the eastern, rain shadow side. Air masses are heated by compression and dried out as they are forced to sink with the prevailing winds down the opposite (leeward) sides of the great ranges, toward the rest of the continent. Extensive open woodlands and sagebrush scrub dominate on these transmontane rain shadow landscapes that stand in stark contrast to those drenched western slopes facing toward the source of the water. The pattern of lush western slopes and arid east sides is a rain shadow effect that repeats throughout California. Many inland valleys even in far northeastern California receive less average annual precipitation than Los Angeles and the Southern California coast more than 700 miles (1,200 km) to the south. Northeastern California locations may be in the

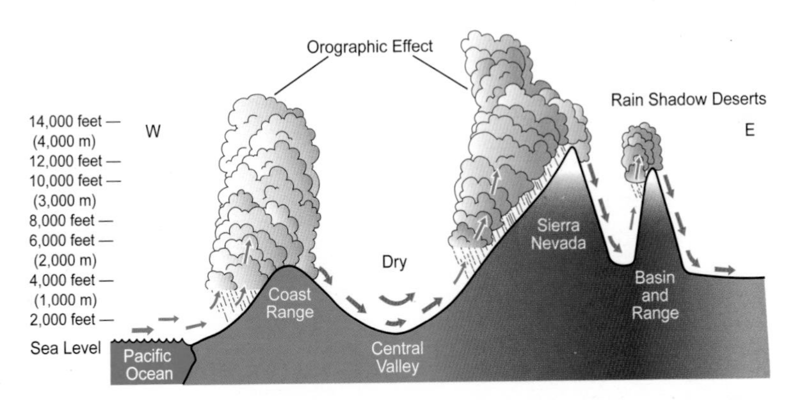

FIGURE 3-6. Orographic precipitation on the windward ocean sides, rain shadows on the leeward inland sides—from Oregon to the Mexican border, we can observe the profound impacts topography and major mountain barriers have on winter air masses, weather, and climates.

direct winter storm path, but they are on the dry transmontane rain shadow side of the mountains; the southwest coast is well south of the direct storm path, but it is on the moist cismontane side of the mountains.

LIFE CYCLES OF CALIFORNIA STORMS

During average years, the jet stream loops farther south as winter progresses. The Aleutian Low strengthens and sags south from Alaska, and upper-level troughs drift southeast, driving storms directly into California from the Pacific. (See the lesson on air pressure and winds in the introduction.) These spinning storms we call mid-latitude cyclones sweep into Northern or Central California, nudging out remnants of resilient high-pressure ridges and dragging high-energy

FIGURE 3-7. This satellite image of water vapor shows a massive Pacific winter-style storm aiming at Northern California. The cyclone is dragging a wet and warmer atmospheric river (whites and greens) from near Hawaii and across the Pacific and then circulating the moisture counterclockwise toward the low, and Northern and Central California. The yellows show drier air. *Source:* NOAA/National Weather Service.

frontal systems across Southern California. Blasts of cold air spin behind these storms, out of the Gulf of Alaska and into far northern California until their blustery winds may briefly drop snow to elevations near sea level, such as at the shores of the Lost Coast about every fifth winter—even San Francisco gets dusted once every few decades.

These massive cyclones are the antitheses to summer's sinking, stable air masses and high pressure. Although each storm has its own unique personality, they share some common characteristics. They are some of the greatest super-scoopers of air and water on our planet, and they are born from winds miles above us in our atmosphere. As the polar-front jet stream sags south and loops around troughs that dig into middle latitudes, high-altitude winds accelerate, creating a void that draws air up from below like a giant vacuum. As air is lifted vertically, less-dense low pressure forms near the surface. (See the diagrams in "How Air Pressure Commands Our Winds and Weather" in the introduction.) Storms are born here (scientists use the term *cyclogenesis*), around the centers of surface lows. These are the largest and most complex storm systems on Earth, and they can grow to more than 1,200 miles (2,000 km) in diameter, far larger than California is long. Enormous masses of air spin counterclockwise into these intensifying centers of low pressure. They resemble gargantuan atmospheric octopi spinning like pinwheels while gradually pulling their arms inward. Contrasting warm and cold air masses meet and clash along boundaries that meteorologists call weather fronts, while the entire spinning cyclone usually moves from west to east.

Let's follow one of winter's circulating monster storms as it drifts off the North Pacific Ocean and into California. As it moves toward the Northern and Central California coasts, warm, moist

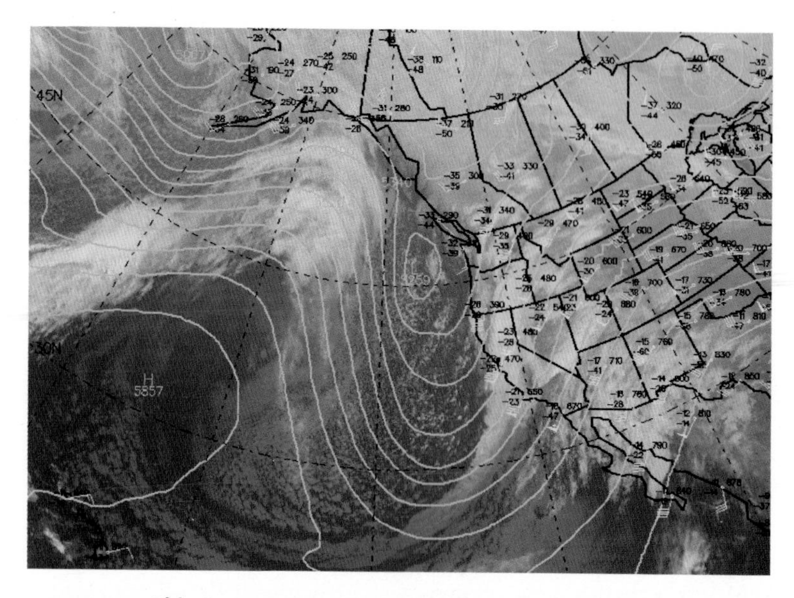

FIGURE 3-8. This January 28, 2021, 500 mb map displays a deepening low-pressure trough digging in the upper atmosphere just off the California coast. Upper-level winds spin and accelerate around the base of the trough just as they are steered over the state, encouraging instability and storminess. *Source:* NOAA/National Weather Service data as displayed by San Francisco State University.

air is pulled up and around (advected into) the cyclone's eastern side from the south and southwest. (Meteorologists often use the term *advection* to describe the horizontal transport of moisture, particles, and energy in our atmosphere.) These warm air parcels from the southern latitudes travel thousands of miles across the Pacific before they can be drawn in (advected) and incorporated into the storm. These air streams, with high specific humidity, transport the moisture and energy that help fuel mid-latitude cyclones. Just as baleen whales scoop up and process tons of plankton and convert them to energy, so do these cyclones scoop up tremendous amounts of water and energy that fuel their engines.

The advancing warm air is forced to glide up and over any denser, cooler air that may have settled around the east (advancing) side of the giant rotating cyclone. As the storm spins closer to the coast, high, thin ice-crystal clouds gradually thicken into cirrostratus and then lower into thicker altostratus clouds. The clouds form still lower until thick nimbostratus clouds bring steady, beneficial light rain and drizzle that can last a full day, depending on how fast the warm front and its parent cyclone are moving. As the warm front gradually creeps by, winds veer from the southeast to the southwest, ushering in more warm and moist air behind the front, but ahead of the larger counterclockwise-churning tempest. Snow levels remain high in the relatively mild temperatures associated with the passing warm front.

FIGURE 3-9. Parhelia (sun dogs) may form on either side of the sun near sunset or sunrise when high, thin cirriform clouds are present. Flat, horizontally aligned ice crystals are refracting sunlight at about 22-degree angles. These cirrostratus clouds may sometimes stream ahead of warm fronts and incoming winter storms.

As the counterclockwise-spinning cyclone comes closer to us, a more volatile cold front is swinging around behind it. Cold air out of the north spins down the backside (western sectors) of the cyclone. The cold, dense, heavy air charges along at the surface, quickly lifting relatively warm air aloft and out of its path. As the warmer, moist air is scooped quickly upward, it condenses, releasing tremendous amounts of latent heat within billowing cumulus clouds along a relatively narrow band. In this turbulent instability, towering cumulonimbus clouds (thunderheads) sometimes erupt, carrying heavy cloudbursts and occasional scattered imbedded thunderstorms with small hail. The drama can build into brief, violent storms that generate local flash floods and longer-term regional flooding when and where soils are already saturated. There are no fine, misty drizzles here; gargantuan raindrops crash and plummet to the surface, making impressive impacts. Keep an eye on your weather vane from inside a cozy shelter, because as the cold front sweeps through, south and southwest winds suddenly veer to the west and northwest, introducing noticeably colder air. These passing cold fronts will demand your attention; the weather changes in abrupt and undeniable ways. This is when mountain landscapes and resorts often experience the cold, fluffy, "dry" snowfalls more typical of climates with bitter cold winters and polar air masses.

After the cold front has passed and moves inland (south and east) with the larger cyclone, we return to tranquility. Imagine the clashing air masses as Godzilla and King Kong (or pick your favorite monsters) fighting—they eventually crash through the city and disappear over the horizon, dragging their battlefronts along, but leaving paths of destruction on the landscape. The battles of the titan air masses leave evidence of their power in the form of wind, flood, and mudslide damage. If you missed the big show, don't worry; there will be plenty of winter storm sequels.

Although all life cycles of mid-latitude cyclones evolve differently, their cold fronts eventually overtake slower-moving warm fronts, merging together to form what are referred to as occluded fronts. These frontal systems may initially spread clouds, winds, and gentler weather normally associated with warm fronts, but they are quickly followed by rough weather we would expect with cold fronts. Look for drizzle and then light rain to start the occluded-front show, which evolves into heavy downpours over several hours. These storm dramas can take a couple of days to spin across the state if they are really slow movers. Winter storms are powerful engines that evolve, transform,

FIGURE 3-10. A powerful cyclone brings fierce wind, heavy rain, and erosive high waves to the Northern California coast. This wet winter storm also caused flooding in surrounding streams and rivers.

FIGURE 3-11. Mammatus clouds hang low ahead of cumulonimbus clouds and cold thunderstorms sweeping off the ocean. Unusually cold air high above has fueled unstable weather; such pop-up storms may build just behind cold fronts or within wraparound moisture on the cold backsides of surface lows.

and reinvent themselves as they move. Air masses, water, and energy flow through and fuel them, but they require those high atmospheric winds to keep the engine running smoothly. After the storms pass to the east and/or south, the unstable cold core of air drifts overhead, igniting pop-up cumulus cloud towers and scattered showers that appear to be chasing the cold front. Look for scattered popcorn clouds on satellite imagery and then in your sky as they billow over the ocean and rotate onshore with the winds that follow the cold front. As pressures build behind the storm, crystal-clear blue skies are decorated with puffy fair-weather cumulus clouds, drifting with refreshing and often blustery northwest winds as the atmosphere stabilizes. I hope that the winter weather patterns I've summarized here will help you better understand and predict weather as you look up at the sky as storms pass through.

Now you are ready to follow winds by looking up into clouds within a storm. When I lived in San Francisco many years ago, I ventured up to Twin Peaks to observe the weather conditions right in the middle of a classic winter storm. The end of the road off Portola Avenue is often flooded with eager tourists clamoring for one of the most spectacular views of the City and the Bay Area. But normal humans were taking shelter from this blustery tempest that produced its own local flooding. A brief lull interrupted the driving rain so that the overcast broke enough to expose layers of clouds streaming in the winds at different altitudes above my location, up through about 25,000 feet (7,600 m). The lower clouds were racing from the south and southeast toward the north and northwest as pressure gradients sucked them *toward* the surface low-pressure center, a storm that would soon come ashore just to the north: winds from the south blowing into the low pressure to the north! But the higher clouds were streaming from the southwest toward the northeast and increasingly from the west, displaying wind shear as I looked vertically through the storm. (Visit the pressure and winds class in the introduction.) Those upper-level winds were blowing nearly parallel to the isobars (imaginary lines connecting points of equal pressure) around the storm. But the lower-level winds and clouds were being pushed and pulled in a tug-of-war that the pressure-gradient force was winning. Surface winds and clouds were streaming from south to north over me as they spun *into* the low in a counterclockwise direction. It is a special treat to watch this wind shear during a big storm, as it requires a brief respite during which there is an opening in the layers of otherwise thick, stacked overcast. I am reminded of a quote from the film *The Boy Who Harnessed the Wind*, which is about a boy in Malawi who builds a windmill:

"Ngati mphepo yofika konse"—"God is as the wind, which touches everything." The winds I observed on that gusty wet day on Twin Peaks are pretty typical of many winter storms. I should also note that upper-level troughs within the looping Rossby waves are rarely vertically stacked directly above surface lows; they are often tilted askew on top of one another like leaning layer cakes. This creates more complicated upper- and lower-level wind patterns that might be better understood with some good weather maps displaying specific conditions at different heights in the atmosphere.

OPTICAL PHENOMENA SIGNAL WEATHER CHANGES

Now you can better understand the old adage "Red sky at morning, sailor's warning; red sky at night, sailor's delight." Really, anyone—not just a sailor—who is planning a day outdoors in California or at a similar middle latitude should watch out for these warning signs. In California, we know that our winter weather patterns are dominated by prevailing winds, air masses, and storms that usually approach from the Pacific Ocean.

When the morning sun rises and shines through clear skies in the east and toward cloudy skies in the west, it can illuminate the bottoms of those clouds in brilliant orange and red. Here in the Golden State during winter, it is likely that those clear skies will continue to move away toward the east, while the cloudy skies (and possible storminess) will be drifting over us from the west. This is because our rainy season's upper-level jet streams, troughs, and their associated storms tend to drift from west to east across the state. Look toward the west to see your future weather. By contrast, when the afternoon sun is setting in clear skies to your west and shining into and under clouds toward your east, illuminating them in red, it is likely that the clouds are drifting away from you and the clear patches will soon drift over you. For the same reasons, another more colorful and perhaps reliable axiom would be, "Rainbow at morning, sailor's warning; rainbow at night, sailor's delight." And by rainbow at "night," I mean just before sunset, when the low sun in the west can shine underneath vertical clouds and into their rain columns toward the east.

You may have sometimes heard the red skies in the adage explained, even by relatively reliable sources, by blaming pollution or dust in the air. It is true that smoke and dust can color the sky red at any time of day, but pollution is not a requirement for producing red sunsets or sunrises. Sunlight turns yellow, then orange and red at sunset because when it is near our horizon, it must shine through a much longer path of denser atmosphere to get to us. Most of our atmosphere's gas molecules (such as nitrogen and oxygen) are small enough to very selectively scatter the shorter visible wavelengths of the electromagnetic spectrum, such as violet and blue. This is what makes a clear, clean, dry sky so blue, and you will find California's bluest skies where there is the least amount of moisture, dust, smoke, and pollution in the air. But when the sun is rising or setting, it must shine through the denser atmosphere near Earth's surface. The violets, indigos, and blues are scattered out first, then the longer wavelengths of greens and yellows, until we are left with the longest visible wavelengths of orange and red as this ROY G BIV cancellation game (follow the letters from right to left) plays out. By the time a low-angle sun gets to the bottom of clouds, only the longest visible wavelengths of oranges and reds remain to illuminate them and then get scattered toward our eyes. When the air is particularly moist and/or polluted, the sun itself may appear yellow and then orange and finally red because there is an abundance of larger water drops or pollution that can finally scatter longer visible wavelengths. The sun and moon also turn red when smoke clouds drift overhead with their relatively larger pollution aerosols.

I've sometimes heard people say that California coast sunsets are red because of smog. This is only occasionally true, however, as it is most often the hazy water drops in the air that will taint the sun red when you are viewing a sunset from the beach. Dominant sea breezes transport clean, relatively pristine air toward the state from the ocean (except for occasional pollution invasions from more distant sources), but that clean air is loaded with moisture fully capable of filtering visible wavelengths until only red may remain to be scattered to your eyes. When this moisture is present, watch the sun set into a deeper red produced by the longest visible wavelengths—until it disappears into the marine layer's thick haze and fog. (You already know to never look directly at the sun when it is higher in the sky; otherwise, it might be the last thing you ever see.)

While old adages and an understanding of cyclonic patterns help us predict storm behavior, there are always exceptions. After some storms drift past us and slow or stall, their spinning winds can fling residual clouds and precipitation back over us in last turbulent gasps before they move on. Some even retrograde back toward us. Every year, I have been tricked by at least one of those rebellious nonconformists. Also, one of those rogue cutoff lows could meander around the state, or how about those summer monsoons that can sneak in from the southeast and off the Sonoran Desert?

FIGURE 3-12. The center of this historic late February 2023 cyclone's counterclockwise circulation has dropped down from Canada to just west of the Bay Area. Surface winds are blowing counterclockwise toward the center of the low. Notice how the extreme northern part of the state is experiencing more intermittent wraparound precipitation now that the record cold storm and its gathering atmospheric river have moved south of that region. *Source:* NOAA/ National Weather Service.

FOLLOWING WINTER STORMS AS THEY HIT THE SLOPES

As storms roll into California off the Pacific, the varying coastal terrain from Cape Mendocino to the Oregon border rises directly into their paths; coastal depressions are warped down between these highlands, creating local variations in cloud formations and precipitation patterns. Around Humboldt and Arcata Bays and the Eel River Delta, the winds from the south and southeast (common as storms approach) leave low-elevation spots around the bays and north-facing slopes within temporary, isolated rain shadows. Then, as storms begin to make landfall, winds often veer to westerly, carrying the rain directly off the ocean and over these locations. Once the main storm or front has swept past, those winds will blow from the northwest and north so that wet and dry slopes also switch. Low-elevation valleys bordered with mountains to the north may now be caught in temporary rain shadows while north-facing slopes will get the brunt of the remaining precipitation. Knowledge of these patterns can help you predict the timing and intensity of very localized stream flooding and higher-elevation snowfall.

As our winter storms move off the ocean and pass across the state, the heavier precipitation caused by air rising along

FIGURE 3-13. Powerful waves crash onshore in Northern California as a drenching Pacific storm makes landfall. Beach foam commonly forms when organic matter in the seawater is agitated by breaking waves, churning up natural froth within these most turbulent, productive coastal waters.

wet mountain slopes becomes a familiar story, and the best term to describe it is *orographic* precipitation. High country in the Klamaths, Marble Mountains, and Trinity Alps Wilderness may be first in the state to be covered with the stunning quiet, crystalline radiance of several feet or meters of snow in a winter wonderland that becomes treacherous and impassable for weeks or months. Subsequent storms may sweep into the Coast, Transverse, or Peninsular Ranges farther south, dumping rain at lower elevations and snow up higher.

When winter's moist, unstable winds approach these ranges at right angles, they can deliver copious amounts of orographic precipitation. Some northern coastal slopes average more than 100 inches (254 cm) of rain each year, while more than 10 inches (25 cm) of rain can fall during any twenty-four-hour period on any coastal slope caught under a streaming atmospheric river. Years ago, these more powerful and far-reaching firehose-like streams of moist air were referred to as the Pineapple Express, among other terms. But these potent air rivers of water flowing across the Pacific don't always originate around Hawaii. That's why meteorologists have settled on the term *atmospheric river* (AR) to describe them. Such deluges have inundated coastal slopes all the way south into the Bay Area (such as near Mt. Tamalpais and Big Basin Redwoods) and Big Sur and more rarely have flooded Southern California slopes. Storm origins and temperatures will determine snow levels, but extreme snowfall events may be measured in feet or even meters per day at higher elevations. The coldest storms may even leave rare temporary dustings on Mt. Diablo above the East Bay, Mt. Tamalpais above the North Bay, at Big Basin in the Santa Cruz Mountains, and even more rarely on lower coastal peaks and slopes that overlook Southern California's urban centers. The spectacular storms and their ARs during the winter of 2023 performed with all of the aforementioned weather extremes. By early spring, seasonal rainfall totals included more than 80 inches around Big Basin, 60 inches at Big Sur, and 40 inches in Berkeley and Santa Barbara. Record snowfall exceeded 700 inches across central Sierra Nevada high country, including Mammoth Lakes. As with other historic AR rainy seasons, the state's reservoirs filled with water, but so did many California farms and neighborhoods.

Plodding through the mud at Marin County's Pt. Reyes and Muir Woods, and Big Basin Redwoods State Park in the Santa Cruz Mountains, a rainy-day hiker can appreciate how plants such as ferns and redwoods have flourished here for thousands of winters, nourished by the dousing each storm brings. But I recommend against visiting when a more powerful AR aims at these mountains. More than just getting soaked, the hillsides and trees become unstable in conditions like those experienced at Mt. Tamalpais in late October 2021. That's when 16 inches of rain fell in two days, thanks to a record-breaking AR that stalled over the coastal mountains and broke many records for one-day October rain totals across the Bay. Similar deluges have been recorded on Big Sur slopes as these firehose water parades sag south and, more rarely, into Southern California's mountains. (More than 10 inches of rain were recorded in one day in the San Gabriel Mountains just above LA during the ARs of 2023.) Even the hardiest individuals must wait for such drought-busting flood events and their mudflows to subside before venturing out.

When I lived in the Bay Area, I would often visit the nearby mountains during and after their winter storms. Dressed in waterproof rain gear, I melted into the quiet serenity of the dripping forests along lonely trails. Like Rachel Carson, I think "a rainy day is the perfect time for a walk in the woods."

WINTER'S COLD AND FOG SETTLE INTO INLAND VALLEYS

After the clouds have cleared and the invasions of frigid air settle behind winter's cold fronts, the coldest nights of the year often set in. Paltry residual heat energy is quickly radiated out through clear skies on long winter nights. Cold, heavy, dense air tends to pool in valleys and other low spots on those clearest calm nights. Severe frosts and freezes in this state that leads the nation in agricultural production threaten billions of dollars of crops, so farmers even in balmy California must make crucial decisions about what they plant where. Crops more sensitive to frosts and freezes (such as citrus, almonds, avocadoes, and specialty crops that include lemon grass) must be protected. Some plants (such as strawberries and grapes) remain tolerant of cold when dormant and can even benefit from freezing temperatures in the dead of winter. But these same plants become extremely vulnerable to late-season spring freezes when they are budding and flowering. You may notice more sensitive species planted where cold air can locally drain past the crops and down the slopes on its way to pool in the lowest depressions, where there are more frequent frosts and freezes. Check out the avocado orchards established on coastal mountain slopes from Santa Barbara and Ventura to San Diego Counties to see where frosts and freezes are least common.

During the coldest winter nights, commercial farmers and home gardeners must take aggressive action to protect their investments. Crops close to the ground may be covered with straw or plastic; gardeners might even use sheets, blankets, or other cloth covers to trap outgoing longwave (heat) radiation. These coverings must be removed as soon as the morning sun

FIGURE 3-14. In this winter scene, stable radiation (tule) fog has settled over wine country. Frost is not a threat now, but these giant fans may be switched on during unusual late-season/early spring cold snaps that can settle around the vines after annual budding has already begun. Different styles of wind-generating blades are designed to break cold inversions near the ground on otherwise calm nights and early mornings.

rises high enough to heat surfaces and break the local shallow inversions, since longer periods of cold into the morning hours do the most damage. Orchard heaters, smudge pots, and bonfires were widely used in the past; they can radiate heat and set off mixing currents in the fields to break freezing temperature inversions near the surface. But burning dirty fuels and waste, such as used tires, is no longer a viable option, as doing so can leave clouds of toxic smoke hanging in a stable and perhaps already polluted shallow layer of air. During these most stagnant conditions, smoke first rises from pollution sources; as it mixes with and cools in the ambient air, it levels off to display anvil shapes that spread out and merge in a layer marking the inversion. Broad, flat smoke "hats" spread over fields and towns. These local inversion patterns prompt air quality districts to declare no-burn days for farming communities. Similar warnings from air quality districts in urban areas have been issued to stop folks from using their fireplaces for heat during stronger inversion periods; someone else's cozy, romantic fire can become your dangerous, choking air pollution.

Wind machines poke up above many Golden State orchards and other farmlands. These popular and most visible investments are switched on when farmers anticipate hours of freezing temperatures that could destroy their crops. The rotating blades can mix the slightly warmer air just above the local inversion into the heavy, dense, coldest air near the surface. When resources are available and visibility is not limited, helicopters have been employed to break shallow freezing inversions. But there are only so many helicopters to act out a commercial farmer's last desperate and expensive attempts to save precious investments. Sprinkler systems are sometimes used to cast water on crops that are about to freeze. Liquid water carries stored heat, and when it begins to freeze, latent heat is released. As thin layers of ice form, they may protect crops by keeping fruit temperatures near freezing when outside air temperatures plummet below freezing.

Just as it takes only a few hours for a catastrophic flood to break through a prolonged drought and cause billions of dollars in damage, so can a few hours of anomalous hard freeze destroy billions of dollars' worth of California crops. For many decades, California farmers have been pushed away from the mildest microclimates on Southern California's coastal plains and into more marginal climates with more frequent frosts and freezes. Displaced farmers lose profits during cold snaps while consumers (including the millions living in developments that displaced the farms) pay more for whatever crops survived the latest freeze.

Although average annual temperatures are on the rise in the long term, more radical swings in weather patterns and extreme weather events are getting locked into place as climates change. Even the most resilient plant communities are struggling to adjust. Regardless, winter's dark cycles of cold and wet will continue to visit the state (at least intermittently) through much of December, January, and February, especially in Northern California.

After dropping some of their moisture, Pacific storms pass over to the backsides of the coastal mountain ranges. Across the inland valleys of Northern California and the Central Valley, the rain shadow effect tends to weaken the storms' impacts, sometimes diminishing the deluge-makers such that they offer only virga (precipitation that evaporates before reaching the ground) and light-to-moderate scattered showers. Still, the meager precipitation leaves enough moisture in inland valleys to enhance winter's shallow radiation fog, known as tule fog or ground fog. The atmosphere stabilizes behind winter storms, causing dense, cold air to drain downhill and settle at the bottom of lowlands from late autumn through winter, especially during the long nights. As overnight temperatures plummet, inversions trap stagnant air masses near the valley floors. The air cools to its dew point in the darkness until the next morning's sunlight drives temperatures a little higher so that the fog is "burned off" into a hazy calm. But the low sun angles of winter don't cast enough radiation to break local valley inversions. Milky haze often waits without a breeze as winter's fleeting sun quickly disappears on the southwest horizon. The

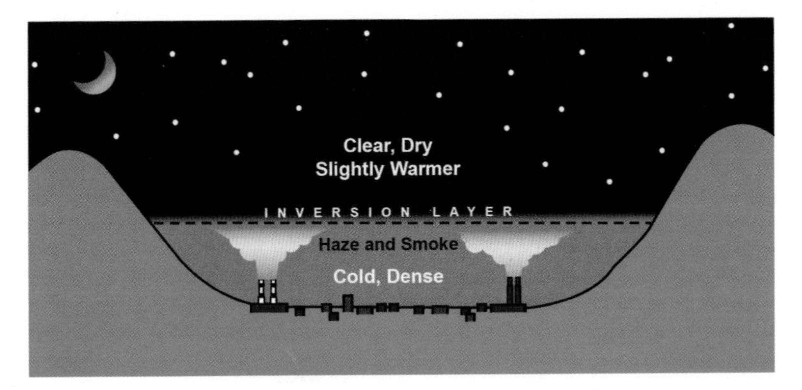

FIGURE 3-15. Local valley inversions commonly form during clear, calm, cold nights in the state's inland valleys.

nearly motionless tule fog thickens in place during each long night without disturbance until visibility diminishes to near zero and air temperatures drop to near freezing. Notice how this radiation fog forms in very different ways at very different locations and times of year compared to advection fog (see chapter 1) that drifts in along the coast during summer.

California's Central Valley is the tule fog capital. We can look down on the shallow cold clouds of cotton blanketing this elongated valley by ascending the foothills until we break through the inversion and into the clear, dry, unlimited-visibility air above. The massive Sierra Nevada erupt to the east and the Coast Ranges emerge to the west out of the murky winter soup below. Saturated stagnant air dominates from the valley floor up to a few hundred to several hundred feet high, while inadequate winter sunlight struggles to evaporate the billions of tiny droplets during each short day. The pool of winter haze and fog may become so dense that it persists through the day and only grows thicker overnight until it can enshroud valley landscapes for weeks; during these episodes, daytime high temperatures barely recover from overnight lows, always remaining near or at dew points.

When widespread upper-level high pressure and stability provide unusually lengthy respites from winter's storms, the damp blanket covers the entire Central Valley from Redding to Bakersfield. It may condense to spread winter's frigid haze into Napa and Sonoma and other wine country valleys. Weak downsloping wind events send dense, dry air cascading from above, only to skim over the cold, stubborn valley fog and toward and over the Coast Ranges. The spreading frigid shallow gloom sometimes creeps into Bay Area locales to condense into a shroud of continuous winter radiation fog with links all the way to the Golden Gate, where tule fog can meet coastal (advection) fog. Such massive undulations of surface cotton are awesome spectacles from above. Bay Area hills and taller buildings may be seen poking up from the cold wintry pea soup. Residents below feel the damp imprisoning chill.

By contrast, mixing air is tule fog's enemy. It could come as a blast of stronger north wind that clears the air around Redding and Red Bluff and scours out the northern end of the Sacramento Valley; or drying winds associated with a Santa Ana pattern, which can occur deep into winter, might swirl into the southern San Joaquin Valley around Bakersfield. The eerie spells are often broken everywhere when winter's turbulent storms sweep off the Pacific, bringing widespread instability and turbulence that mixes air columns, destroys inversion layers, and disorganizes and dissipates fog banks. These cycles repeat themselves at intervals throughout the late fall and winter, depending on whether it is an active wet year of constant mixing charged by transient weather systems, or a stagnant-weather year. And although the Central Valley's tule fog is legendary, similar radiation fogs envelop smaller California inland or mountain valleys during winter. They also condense in the cold, stable air that settles in low spots behind wet storms, and they are frequently found in frigid stagnant air just above ice- and snow-covered depressions surrounding mountain resorts. Surface heating, high winds, or invading air masses brought by the next winter storms will eventually mix out and evaporate them.

Shrouds of radiation fog cast deadly spells each year when motorists think they can drive just as fast as ever, even when visibility deteriorates down to a few feet. It is a foggy nightmare that becomes deadly in a second for each driver and passenger, as casualties accumulate over several long minutes until the worst accidents have included hundreds of vehicles and scores of victims. When the fog clears, the highway tragedy resembles

FIGURE 3-16. Winter's extensive tule fog has spread beyond the Central Valley, into inland valleys around the Bay Area, until it forms over the Bay around San Francisco. The Golden Gate and San Francisco rise above the shallow soup that enshrouds the Bay Bridge in the distance. It may be burning off during the afternoon, only to condense again overnight should the cold air mass remain stagnant.

a bizarre scene from Land of the Giants, as if colossal kids had been tossing and crashing their toy vehicles without supervision. Instead of tempting fate and attempting to race through the fog, both literally and figuratively slow down or safely turn off your real or imagined paths, and establish a safe distance from the frantic traffic. Take some deep breaths and find your emotional refuge in the dank, silent curtains of minuscule water droplets that obscure your larger view of a frenetic world. Rediscover your peace in the cover of nature's protective womb of moisture. Smell the accumulated scents of dampened leaves, soil, and other natural materials and the various air pollutants that may have been added into the atmospheric soup recipe by nearby vehicles, farms, industries, and other human sources. Appreciate how the isolated remnants of natural communities (such as what remains of the valley's prairies and riparian woodlands) will benefit from the cool moisture caught by plants and

FIGURE 3-17. Nimbostratus clouds thicken as a Pacific storm approaches the San Jacinto Mountains. An undulating layer of some combination of radiation and upslope fog is trapped at the base of the mountains surrounding Garner Valley.

soils that will bake and dehydrate in a few months within these dry inland valleys. The surrounding calming dormancy will not last long before the spring germination and budding season erupts out of the fleeting life-giving moisture.

As land surfaces are heated during longer days with more intense sunlight, the cyclical inland radiation fog eventually evaporates during spring and disappears for the summer. And now you know why the summer coastal fog described in chapter 1 couldn't be more different than the fog we explored here.

WINTER'S STORMS ENCOUNTER OUR HIGHEST AND LARGEST WALLS

Let's follow the storms as they move to the east of the inland valleys. These same cyclones that drenched the coastal mountains will eventually encounter California's loftiest mountain barriers: the southern Cascades and the mighty Sierra Nevada. Although the state's northern volcanoes exhibit locally impressive wet windward and dry rain shadow slopes, the Sierra Nevada display one of the most formidable precipitation dams on the planet. The mountains' lofty crests line up as almost perpendicular barriers to some of the wettest west winds when winter's storms pass through. As the air is gradually lifted up and along relatively gentle western slopes, cooler storms can dump several feet or even meters of snow in one day. Warmer storms, often associated with atmospheric rivers streaming from the subtropics, are capable of delivering prolonged downpours that can quickly melt snowpacks, send torrents of water in to flood canyons, and suddenly fill reservoirs to capacity. Many crippling record droughts and desperate water shortages were broken by

ARs that have plunged the state into flood crises within a few days (such as in 2017 and 2023), thanks to these mountains that become giant water catchers. So much water is drained out of these Pacific storms as air rises up western mountain slopes that average annual precipitation actually *decreases* as we ascend up to the very highest elevations, above the tree line. But because almost all precipitation falls as snow and temperatures remain colder, snowdrifts last longer at these loftiest elevations; their lingering frozen patches lead to the misconception that the very highest peaks receive the greatest amounts of precipitation. If you hike the high country and notice a patch of snow remaining well into summer, chances are good that it was a massive drift blown and deposited during winter's whiteout blizzards into a shady, sheltered natural amphitheater.

FIGURE 3-18. Giant, relatively "warm" mushy snowflakes add to the snowpack in conifer forests on Sierra Nevada slopes just above Lake Tahoe.

It is important to reaffirm that each storm has its own personality and peculiarities, but there are always some foundational similarities. Rain and high snow levels first dominate as the warm sector spins up on the approaching east side of the mother low (mid-latitude cyclone). As the storm center approaches, higher-altitude temperatures plummet. The bitter-cold ice crystals and snowflakes falling from these cold, unstable clouds take longer to warm to their melting points, so snow can frequently fall at elevations remaining a few degrees above freezing. As the low-pressure center drifts overhead, colder air circulates down to the surface behind any passing cold fronts. This may briefly plunge snow levels into the foothills, while frigid higher elevations are buried in cold, fluffy, "dry" powder. By now, moist air masses with higher water content have been pushed away to the

FIGURE 3-19. This Tahoe winter scene shows how mountain architecture is designed to withstand wet, heavy snows that can accumulate to several feet in a day.

east and south as colder air barges in. As the cold backside of the storm becomes moisture starved, the clouds and steady precipitation break, leaving only scattered residual snow squalls and flurries drifting from west to east and/or north to south up and along the mountain slopes. Emerging animals and people are forced to modify their routines to navigate through their bright new world.

Even without measurable precipitation, drifting stratocumulus clouds containing supercooled water droplets (liquid drops with below-freezing temperatures) sometimes ascend along high mountain slopes while, or just after, cold fronts pass through; forests may be enshrouded in fog where surface temperatures and exposed objects remain below freezing. The droplets freeze directly onto those surfaces, growing into milky granular shapes to create winter wonderlands of rime ice growing on trees and other objects. Should the drifting moisture encounter frozen bridges and roads, it can accumulate as slick, dangerous black ice. In addition, when cold, clear air is near its dew point and it comes in contact with surfaces below freezing, hoar frost forms, as water vapor from the air freezes directly into ice (a process commonly called deposition); delicate hair-like ice crystals grow on exposed surfaces. Nature has her own ways of flocking holiday trees and other objects, but you'll have to travel up into the mountains to catch these frigid winter-wonderland-in-California spectacles.

Air masses often stabilize behind these winter storms as they and their cold fronts march on toward the Rocky Mountains, leaving crystal-clear skies decorated by scattered fair-weather cumulus clouds. Winter's weak sunlight and strong winds help dislodge clumps of snow that break out of trees and plop into the snow blankets. The snow seems to absorb every sound, creating a peaceful silence. By early spring, the layers of wet and dry Klamath, Cascades, and Sierra Nevada snowfalls often accumulate to more than 10 feet (about 3 m) deep above 6,000 feet (1,830 m) and up to 20 feet deep at choice locations during wet years. Individual record snowstorms include the 189 inches (480 cm) from one storm at Mount Shasta Ski Bowl in 1959. Multiple snow records have been set near Tamarack along the Alpine/Calaveras county line (such as 32.5 feet [9.9 m] in one month, January 1911) and at other Sierra Nevada locations. Donner Summit is often the regional winner with average snowfall totaling more than 400 inches (30–40 feet, or 9–12 m) each year. Successive storms during wettest years have dumped up to 70 feet (21 m). The 2023 ARs delivered more than 700 inches (18 m) to central Sierra locales, setting several all-time records. Such epic snowfalls became legendary after the Donner Party was trapped in 1846 and lost about half their members. Later, snow tunnels and protective sheds would be engineered so that trains could navigate Sierra Nevada passes during most of the winter.

Today's wettest winter storms leave several feet or even meters of fresh blowing snow to block transportation corridors and briefly interrupt mountain life and businesses, including ski resort operations. The top of Mammoth Mountain becomes dangerous during these blizzards. Resort ski patrols use early morning explosives to clear the way for safe snow sports while skiers and snowboarders are sleeping off their hangovers. In dawn's early light, listen for booming and rumbling sounds on distant slopes that advertise awesome ski conditions for the new day. These piles of snow keep Mammoth and Tahoe ski resorts open into July in the wettest years, when snow plows often struggle to reopen Tioga Pass by June or the main road through Lassen Volcanic National Park before July 4. (I have walked at the base of melting snow drifts several feet over my head next to a partially frozen Lake Helen in Lassen Volcanic National Park in early July, following one of California's wettest El Niño years.)

ADJUSTING TO ERRATIC SNOW SEASONS

Winter in the high country is slumber time for black bears. They make their beds beneath the protection of rock formations, downed logs, piles of debris, or any place that suits them. Some (such as around Mammoth) break into warmer crawl spaces below cabins and other structures. Other animals shelter from the deadly winter winds in their burrows. Traveling fauna have already migrated to points south or lower elevations or both. You will find some tracks of predators and prey that remain active in the snow. Snowshoe hares will eat twigs and bark, while carnivores such as weasels feed on rodents and rabbits. Both the hares (prey) and the weasels (predator) blend in with their white winter camouflage fur. Looking down on this activity are isolated stands of dormant aspen, but conifers' needle-like leaves are well adapted to the cold so that the trees are ready to carry out photosynthesis during any sunny warm spell. This is why you will find trees such as lodgepole pine and fir growing into high elevations. You can also find yellow-green staghorn lichens growing on the trunks of these trees down to the average annual snow height.

Unfortunately, over the last few decades climate change has also orchestrated winters when little or no natural snow fell, much less accumulated during the ski season. In these cases, machine-made snow is required to produce slender strips of ski runs to serve more avid skiers and snowboarders and those who made vacation reservations months ahead. High-pressure pumps spray mist into the below-freezing air to make "fake" snow that accumulates on surfaces. More than one California ski resort has survived climate change's erratic, unpredictable winters by relying on these snowmaking schemes. Natural plant communities and their animals are not so lucky.

By late December, successive days below freezing and hard-freeze nights transform high-country mountain lakes into ice, especially in the Sierra Nevada and Northern California. The freezes are so reliable that before the days of widespread refrigeration, block ice was cut and harvested from some of the more accessible frozen water bodies. One particular lake experience stands out for me from about a decade ago. My family and I

FIGURE 3-20. You would expect to find several feet or even meters of snow accumulating at the top of Mammoth Mountain and nearby peaks in late December. But not this year. Instead, rocks poke up from thin layers of mostly "artificial" snow, blown out of snow cannons, to keep at least a few ski runs open.

made reservations at Mammoth Lakes for the end of December. We were met with exposed surfaces and little or no powder that year. Because temperatures were well below freezing, we enjoyed walking on frozen Mammoth Lakes and other ice-covered lakes in the Sierra Nevada. A few locals were even skating and sledding on the lakes. Standing quietly, we could hear the mysterious pinging, popping, and booming sounds made as cracks propagated and sent their reverberations through the restless ice. Caution is always warranted when stepping out on the ice—unwary tourists (and some would-be rescuers) have plunged through to their hypothermic deaths in these frozen-surface fairylands.

Seeing bare scree during a Sierra Nevada winter strikes fear into the hearts of mountain folks and flatlanders alike. The moisture content of the annual snowpack determines how

FIGURE 3-21. Julie, Jim, and I and other visitors walk, skate, and sled across frozen Mammoth Lakes. The ice was typically plenty thick and safe to support us during this late December day, thanks to nighttime temperatures that can plunge to near 0°F and days that may not make it above freezing.

FIGURE 3-22. Fed mostly by snowmelt, the Truckee River is the only outlet for Lake Tahoe. This precious meltwater nurtures miles of Sierra Nevada ecosystems in California (home to reintroduced beavers) before taking an abrupt turn east toward Nevada. That eventual turn distinguishes it from most other Sierra Nevada rivers.

much water will fill reservoirs during spring and summer so that it can then be distributed to farms and cities during the long Mediterranean high-sun drought season. The orientation, height, and profile of the Sierra Nevada combine to create perfect water catch basins because heavy orographic precipitation and then runoff accumulate along the extensive and relatively gentle western slopes. The runoff then conveniently flows down those slopes toward the Central Valley and into reservoirs and water projects for storage and redistribution. As the winter season progresses toward spring, hydrologists must frequently measure the snowpack to determine what combination of flood control and water storage must be emphasized. Miscalculations can result in downstream death and destruction and billions of dollars in losses, especially in agriculture.

Core samples are used to measure the depth and weight of the annual snowpack. (The highly variable average ratio of snow depth to water content is 10:1.) Heavy "wet" snowflakes falling near their melting temperature are malleable and easily squashed together, forcing out air pockets; colder, "drier" snowflakes and ice crystals stack more rigidly, trapping numerous larger air spaces between. Such lighter, fluffier powder preferred by skiers and snowboarders is more common at colder inland resorts such as in the Rocky Mountains. This "dry" powder doesn't contain nearly as much water as the warmer, "wetter," heavier snow (sometimes called "Sierra cement"), which is notorious for collapsing roofs and is more common to milder air masses flowing directly off the Pacific Ocean and into California. Each year, these measurements provide critical data to the hydrologists and engineers who will take responsibility for saving water for drought or releasing it to guard against the coming floods.

AVERTING THE DEADLIEST "NATURAL" DISASTER IN CALIFORNIA HISTORY

The near failure and destruction of the Oroville Dam spillways in early February 2017 is one extreme example of how nature's weather cycles can evolve into life-and-death lessons. A series of ARs suddenly began filling Northern California reservoirs that had been drained nearly empty by five years of unprecedented drought. The state in severe drought mode was suddenly thrust into flood disaster control mode. Oroville's spillway was employed to drain water that was accumulating in the reservoir faster than it could be shed. As torrents of cascading water eroded gaping ravines in the spillway, Oroville's emergency spillway was utilized for the first time in the dam's history. But this quickly caused the foundation rocks below the emergency spillway to also rapidly erode, compromising the integrity of the entire system.

During a professional conference field trip, we talked with local public officials (including the county sheriff) who during the emergency got an unexpected crash course in meteorology, hydrology, and engineering that lasted for a couple of days. They told us how a sobering emergency event suddenly evolved into a terrifying crisis. More than 180,000 people were forced to evacuate downstream (in what may have been the largest evacuation with such a short warning in US history), fleeing a potential 30-foot (10 m) wall of racing water. At the final hour of desperation, engineers and hydrologists were able to find just the right mix of releases to keep the dam's systems from failing, averting what may have been the worst "natural" disaster in California history. You could sense the entire state breathing a collective sigh of relief. More than $1 billion was later spent to repair and fortify Oroville Dam's spillways, reminding us that investments in upkeep of infrastructure protect us from potential weather disasters that could cost many billions of dollars and hundreds or even thousands of lives. Closer to the coast, thousands of residents have also learned the importance of long-term investments when they were inundated in the Pajaro River flood of 2023 and several historic floods before that. For years, hydrologists had warned that the inadequate levee system would be breached again, but too little was done to protect entire farms and neighborhoods that were flooded out when the 2023 winter storms caused a levee failure.

FARTHER EAST, STUCK IN THE RAIN SHADOW AGAIN

Once their wet air masses have reached the mountain crests, winter storms are free to cascade down those steep eastern slopes that are being lifted nearly vertically along active faults. Already drained of abundant moisture, the descending air is heated by compression and dried. By the time these prevailing winds and storms reach the Owens and other desert valleys on the continental side, they are often spent, stabilized, and dried out. The Sierra Nevada blizzards often dissipate into fallstreaks (falling ice crystals that evaporate before reaching the surface), virga, and partly cloudy skies as they pass over the Owens Valley. The next orographic boosts will be in lesser but nearly as lofty Basin and Range mountains such as the Whites, Inyos, and Panamints to the east. But these dry ranges are barricaded behind the behemoth Sierra Nevada wall that has already captured those massive snowpacks. Winter's thick white blankets on the western slopes of the great "range of light" (as John Muir called the Sierra Nevada) gradually melt to support the largest trees in the world, giant sequoias (*Sequoiadendron giganteum*). But you won't find such moisture or those kinds of forests in the rain shadow ranges downwind and to the east of the Sierra Nevada.

What you *will* find are wild extremes that define the weather in these dry, windy, harsh transmontane climates that straddle and then extend across Nevada. Here is where higher-elevation temperatures can freefall below freezing on any summer night and drop well below zero on any winter night, as longwave radiation quickly escapes through the clear, rarified atmosphere. As an example, the high-desert ghost town of Bodie is renowned for recording the nation's coldest overnight low temperatures

FIGURE 3-23. A winter storm off the Pacific dumps snow squalls on Sierra Nevada slopes in the background, but the precipitation can't make it down into the dry rain shadow of the Alabama Hills in the foreground.

outside of Alaska, often on summer nights; frigid winter winds will cut right through you during most of the year. Except for a few summer months, almost all precipitation falls as snow at highest elevations in transmontane California, but dustings of cold, powdery "dry" snows are more common compared to the great Sierra Nevada wet-snow dumping grounds to the west. Fierce winds can drive blizzards of blowing snow along the ground until it accumulates and rests in drifts within sheltered amphitheaters. The drifts disappear slowly at higher elevations facing northeast and under the shadows of rock formations, until they survive only as small patches of winter's frosty memories melting into the summer.

These harsh weather conditions torture any living organism calling the Basin and Range its home. Few species can survive in the dry, salty basins that turn into frying pans during summer and then into shallow atmospheric pools where subzero-Fahrenheit air settles on calm winter nights, followed by desiccating winds that frequently whip dust clouds into swirling frenzies. Hardy sagebrush scrub species adapt to extremes on the xeric alluvial fans, bajadas, and mountain slopes and then mix with pinyon and juniper woodlands at higher elevations. On still higher subalpine slopes, scattered limber pine (*Pinus flexilis*) and Great Basin bristlecone pine (*Pinus longaeva*) pop up through rock formations near the timberline. Bristlecone pines are not only the oldest known individual trees in the world (the oldest dated at just over 5,000 years in California's White Mountains), but they look the part. These twisted, gnarly antiques grow painstakingly slowly in nutrient-poor dolomite soils.

Local temperatures can plummet to −40°F in winter within some colder regions of transmontane California. (At barely over 5,500 feet elevation, Boca, just north of Lake Tahoe, recorded the state's long-standing coldest temperature at −45°F [−43°C] on January 20, 1937. No wonder this was an ice-harvesting site during the years before refrigeration.) As in the Sierra Nevada and other highest elevations in the Golden State, plants are often dwarfed to the height of the annual insulating snow cover and below the occasional snow avalanche that may tumble down steeper slopes. The term *krummholz* is used to describe how they are twisted and sheared as they grow, sculpted by the prevailing winds of torment that blow shards of ice resembling tiny pieces of broken glass. Labels such as "banner" and "flag" describe how you can guess the direction of the strongest prevailing winds, as the trees grow pointing downwind. Winter in these ranges can become an unrecognizable and brutal perversion of the mild weather that helped make California legend. And you

FIGURE 3-24. Flag trees near Lassen Peak. In this scene, they are sculpted by winds blowing around and over the peaks from right to left.

can bet that when the Mamas & the Papas recorded their fabled "California Dreamin'," they weren't singing about these parts of the Golden State on such a winter's day.

At times, often between winter storms, a strong west-to-east weather pattern (meteorologists use the term *zonal wind*) pushes relatively stable air masses within high-velocity westerly winds (around 46 mph or more) up and across the Sierra Nevada. When these winds meet the mountains at right angles, they are squeezed directly up and over the crests, similar to a heavy boulder rolled to the top of a hill; but the air parcels are too heavy and stable to continue rising higher above the mountains. Instead, once they clear the crests, they roll downward, pulled by gravity, toward their preferred lower altitudes on the eastern side of the range. Momentum leads to the formation of mountain waves (locally known as Sierra waves) on the leeward side of the range. Imagine that the air molecules behave like passengers on your favorite theme park roller coaster. They are gradually pushed and pulled up the western slopes until they reach the very top of the ride, where there is a momentary sense of weightlessness. Gravity will suddenly send them freefalling down the steep eastern slopes (producing that nauseating, free-floating, plunging feeling for the human riders). Momentum then carries the passengers bobbing up and down over the undulating tracks that extend downstream beyond that initial highest heart stopper. Likewise, such oscillating lee waves can carry far downwind as they meander up and down on their way east, always seeking equilibrium. The downstream waves are enhanced by north–south-trending Basin and Range topography until they tumble and cascade all the way across Nevada. Giant vertical rotor (circulating) wind motions often spin above the deeper basins framed by mountains on each side, occasionally strong enough to kick up dust from the desert floor. Some similarities to the Washoe Zephyr winds of summer might remind you of the connections between different weather patterns during different seasons.

When there is sufficient moisture, stationary lenticular (lens-shaped) clouds will form as wind meanders upward, expanding and cooling to its dew point. Lower stratocumulus and higher altocumulus lenticularis clouds or even higher cirrus clouds grow as small patches and rows at the crests of the waves. Stationary lens-shaped banner clouds also appear as caps above individual peaks (such as Mt. Shasta) as moist, stable air streams up and over isolated mountaintops. When there is more moisture, larger pancake- or flying saucer–shaped clouds with peculiar layers condense above and downwind of major ranges, sometimes linked in rows, but always remaining relatively stationary as the wind blows through them. The wind loops up to condense on the windward sides of the wave crests and slides down to evaporate on the leeward sides. In these cases, pilots and gliders can see the turbulence ahead and anticipate where the updrafts and downdrafts are located. Those turbulent roll eddies, or rotors, often circulate just below the wave crests and may spin vertical rotor clouds when the air is moist. Lenticular cloud formations can also condense on the coastal sides of mountain barriers within more rare offshore winds, casting leeward waves toward the ocean.

Back on that eastern edge of California, the rain shadow effect occasionally conjures strange weather disappearing acts, as I have observed above Death Valley. When potent winter storms are soaking coastal areas and dumping tons of snow on Golden State mountains, a giant oval hole often appears in the clouds over this deepest desert basin, allowing the sun to illuminate the salty flats and sinuous sand dunes. You can see the fuzzy streaks of rain and snow dangling from clouds, like giant jellyfish, as air must rise over nearby mountains, dumping moisture on the higher surrounding slopes. But the same clouds evaporate as air sinks downslope toward the valley floor. This explains the annual average of just more than 2 inches (5 cm) of rain, yet another reason why Death Valley lives up to its name. The same patterns repeat themselves in the region's isolated desert basins near the Nevada border, such as in Panamint, Saline, Eureka, and Fish Lake Valleys. (A little place called Bagdad in the rain-shadowed Mojave Desert holds the state record for the longest drought with no measurable precipitation: 767 consecutive days, from October 1912 to November 1914.)

Another extreme rain shadow demonstration took place when ARs streamed into central California during the winter of 2023. Directly to the west, some coastal slopes accumulated more than 80 inches of orographically enhanced rainfall from those winter storms, while record snows buried the Sierra Nevada. Less than 300 miles from the coast, but a few mountain barriers downwind (east), lowly Death Valley recorded less than 1 inch during the same period! The no-rain zone extended throughout the lower deserts all the way down to the Imperial Valley and the Mexican border. I could almost sense the disappointment among the dehydrated mesquite, creosote, desert holly, kangaroo rats, coyote, and other plants and animals. Surrounded by their dry, salty and sandy soils, they could only look up and to the west to see the record rains of 2023 that never came down. They not only had to survive the freezing, windy winter nights but wait and hope for summer's monsoon.

WINTER STORMS ARRIVE IN SOUTHERN CALIFORNIA

When winter storms bear down on Southern California, the region's impressive mountain ranges impact weather patterns in ways that might remind you of the mighty Sierra Nevada. But there's a big difference: Southern California storms encounter the Transverse Ranges, which trend west–east across the region. These nonconforming walls of rock have been rotated, faulted, and folded along the big bend in the San Andreas Fault system. Follow them on the map from west to east as they rise above the ocean at Pt. Conception and Pt. Arguello, along the Santa Ynez and other individual mountain ranges, all the way to the lofty San Gabriel and San Bernardino Mountains, which peak at 11,500 feet (3,505 m) at their eastern extent, before they descend toward Joshua Tree National Park. Not only do they prevent cold snaps from the north from reaching Southern California's coastal plains in the LA Basin, much of Orange County, and the Inland Empire, but the Transverse Ranges play a vital role in determining where and when precipitation will fall when winter storms finally sweep through Southern California. Because the jet stream steers storms off the North Pacific, those storms most often sweep across Northern California first. Some larger cyclones eventually push frontal systems into Southern California. In other, rarer cases, upper-level troughs may dig farther south so that Central and/or Southern California receives direct hits. We will follow one of these storms as it interacts with the formidable Transverse Ranges.

As with other winter storms, winds blow counterclockwise into the low center. Although the cyclone may be approaching from the northwest, it will be pulling warm, moist air up from the southeast, south, or southwest. Trees and flags point toward the north in the southerly, backing winds as our giant super-scooper low-pressure suction system drifts closer. This south wind might also trick less savvy observers into thinking that the storm is approaching from the south. Various layers of mainly stratus clouds thicken as warm, moist air is drawn into the storm. Since the upper-level trough and the center of low pressure are still far away, there is no trigger to destabilize the atmosphere and encourage turbulent, rapidly rising air that we call convection, which might otherwise build cumulus clouds and thunderheads. Such relative stability often results in a sky full of lowering and darkening clouds but mostly flat, innocuous ones (cirrostratus and altostratus) until lower stratus and stratocumulus clouds begin streaming overhead, toward the north or northwest. Long trains of subtropical moisture may be dragged out of the tropics and across the Pacific to be incorporated into the larger circulation. Steady drizzle and light precipitation become widespread along with other weather conditions associated with warm fronts.

As the cyclone gets closer, winds increase and veer slightly until they flow from the south and southwest, transporting relatively warm moisture and nimbostratus clouds with them. Sound familiar? But this time, topographic barriers force the air to rise, cool, and condense into thicker clouds, releasing rain onto south-facing (windward) slopes. As the moist air ascends uphill toward the top of the range, the rain and snowfall become heavier. As the storm drifts closer with its cold core of air, snow levels drop and rain gradually turns to snow at lower elevations. At mountain locations such as Wrightwood and Big Bear resorts, one storm can deliver a quarter or more of the average total precipitation for an entire year.

This orographic lift explains how these mountain slopes average more than twice as much annual precipitation as the crowded coastal flatlands below. Epic floods (and deadly debris flows, as discussed in chapter 1) have visited the base of these mountains during historic storms. When Los Angeles measured nearly 6 inches of rain in one day on March 2, 1938, and 11 inches in a five-day period, at least 30 inches were measured in the mountains that ring the basin. Those short-lived but horrific 1938 floods killed more than a hundred people, left much of LA, Orange County, and Inland Empire flatlands under water, and incentivized years of costly flood control infrastructure projects that included lining water channels in concrete all the way to the beach. This overresponse left ribbons of paved river channels looking like massive flood-control ditches running across the LA. Basin. Costly projects to rip up the concrete, reclaim the rivers, and restore riparian plant communities are under way along parts of those channels. Today's sprawling catch-and-spread basins at the base of the mountains are designed to pool floodwaters and encourage infiltration to recharge local groundwater supplies.

Plant communities change from coastal sage scrub and chaparral to woodlands to conifer forests as you travel farther up toward cooler and wetter higher elevations of the San Gabriel and San Bernardino Mountains, until you reach the very highest subalpine and tree-line mountaintops. (Refer to the vegetation zones diagram in chapter 1, figure 1-20.) Threats to downstream populations have encouraged smarter management and nurturing of these wildlands and their plant communities that provide us with inexpensive, natural flood control. Wildfire-to-flood cycles will keep reminding us.

Once the rising air scales the great barriers' peaks and crests and spends much of its moisture, it is free to slide downhill along the northern rain shadow slopes and into the high desert, transformed into drier and more stable air masses. If you followed the winds as they descend down the backside of the mountains and into the desert, you'd see how the surrounding forest thins to pinyon and juniper woodlands and then Joshua tree woodlands and desert scrub. The Antelope Valley and familiar cities such as Palmdale, Apple Valley, and Lancaster are considered to be in true deserts because of this rain shadow effect, while you will find a moister Mediterranean climate on the wet (south) sides of these ranges facing the ocean.

The Palmdale wave is strikingly similar to the previously reviewed Sierra wave. But here, high-velocity winds drive a relatively stable air mass from south to north, or southwest to northeast, roughly perpendicular to the Transverse Ranges. Once forced up and over the top of the mountains, the winds can blast down the desert side and into the Antelope Valley. Momentum causes the winds to hurtle below their equilibrium levels as they dip toward the surface; such high winds have created havoc and destruction on the desert floor. Similar to what they do on the

FIGURE 3-25. A winter storm has just passed, leaving a bank of cold clouds hugging thick blankets of snow in the San Gabriel Mountains. The palm trees near sea level and the not-so-distant mountain snow remind us that it is still possible to surf and ski on the same day.

east side of the Sierra Nevada, they compensate by looping back up into the lee wave, a process that may repeat itself as the otherwise relatively stable air mass continues its roller-coaster-like wavy looping and vertical meandering downwind and across the desert. Look for the dust storms where these oscillating winds swirl over disturbed construction sites and dry lake beds. When there is moisture, you might spot some of those very impressive lenticular clouds and rotor circulations (caused by vertically rotating eddies) that are also similar to those found on the downwind sides of the eastern Sierra Nevada.

This winter storm also brings dramatic changes as it moves nearby or overhead. As the characteristic cold front circulates around the back side of the larger cyclone, warmer air will finally be forcefully scooped up along a relatively steep and narrow cold front. Vertically developed clouds bring cloudbursts and even thunderstorms. As the turbulent cold front passes through, gusty winds will quickly veer to flow out of the west and then northwest, ushering in dramatically colder air on the backside of the storm. Even the foothills will experience snow as colder air invades behind the front. This is when you know that the front and the heaviest precipitation and storminess have passed through. Watch as the well-defined cold front sweeps through with its wall of ominous dark sky towers, and then look toward the northwest to find the first breaks in the clouds. If the front is followed by building high pressure, a steep pressure gradient may propel strong, blustery winds that veer from westerly to northerly over time.

With such a wind switch, the same Transverse Ranges act as barriers that now catch residual clouds and precipitation racing out of the *north* as the gusts chase the storm south and east. This is when clouds and showers may bank up against north-facing slopes, while regions on the shielded south (now leeward) side of the Transverse Ranges clear out, in an end-of-storm reversal of the winds and weather conditions that started this commotion. As the north slopes of the Tehachapis catch some of this precipitation, busy Interstate 5 may become treacherous or even blocked at the top of the Grapevine by snow flurries, ice, and freezing fog.

Wildlands Conservancy's Wind Wolves Preserve may be one of the best places to watch the effects of topography as veering winds mark the passage of winter storms. The preserve is located on north-facing slopes of the San Emigdio Mountains near where the Tehachapi Mountains meet the Transverse and Coast Ranges. The preserve's prairie (grassland) slopes rise out of the San Joaquin Valley until they transition into higher elevations of oak woodlands, pinyon and juniper, and eventually cooler and wetter plant communities with ponderosa pine and big cone spruce. Because they face north, these slopes often experience chilling residual showers in the northwesterly or northerly winds that follow on the heels of winter storms. When much of Southern California may be clearing after a storm, Wind Wolves' slopes can be enshrouded in clouds and showers that mark a storm's grand (or fizzled-out) finale. When my students and I were at the preserve, we were able to sense these unique air currents and weather patterns where we camped and helped clear out invasive nonnative species from the native plant communities there. If you can't visit these striking landscapes, you might notice on satellite images the clumps of postfrontal clouds that frequently bank against the slopes in the north and northwest winds. This cusp of merging landforms rising at the horseshoe-shaped southern end of the San Joaquin Valley resembles a huge lifted amphitheater that can trap and block these north winds until their moisture is wrung out.

Weaker storms and frontal systems approaching from the northwest or north may fizzle out when they are forced to make the turn (forecasters may refer to it as "turning the corner") around Pt. Conception and Pt. Arguello; they are often debilitated and disorganized in their attempts to penetrate through or detour around the Transverse Ranges. Regardless, assuming that the supporting high-level winds and troughs continue south, east, and inland, pushing the cyclone along, the embattled front will likely further dissipate as it moves into Nevada or Arizona or toward the Mexican border.

Before we track Pacific storms even farther south, we should be on the lookout for atmospheric surprises that can suddenly change forecasts in Northern or Southern California. When upper-level troughs slow or stall on approach, the original cyclones, fronts, and other near-surface features might continue ahead of them and lose their support aloft. In these cases, the cold-front passage may not be as dramatic, and the winds may even remain rather calm in the weak pressure gradients and clearing skies behind the storm. You could be in what's known as a sucker hole. This occurs when you think the turbulence and inclement weather has passed with the front, but a low-pressure trough still looms off the coast or upwind. As long as the cold core of air in the upper trough remains upstream or overhead, extreme temperature gradients will exist between the air near the ground and the very cold air above. The differences between air temperatures near the surface and aloft are known as lapse rates. When these lapse rates are extreme (or steep), they will encourage instability and cyclogenesis (conditions where storms can form and strengthen). A series of smaller storms, often carrying colder, unstable showers that may include imbedded thunderstorms and hail, may sweep into the state. The bright popcorn clouds you see on satellite images are observed from the ground as towering puffy cumulus clouds. Such instability has even generated rare water spouts and rarer tornadoes when the cold upper cores of air drift over flat regions such as the Central Valley or the LA Basin. After the afternoon sun has had a chance to heat the surface, amplifying the difference between the relatively warm air near the surface and the very cold air aloft, the sky's mood gets even more unstable. While always rare, funnel clouds and tornadoes do tend to peak later in winter and spring (in cismontane California) during the afternoons when more intense sunlight has broken through to heat the surface during or immediately after a storm's passage.

There are other exceptions that make these dynamic winter storms more complicated. Perhaps most notable is when elongated and narrowing southern loops are carved into upper-level troughs until they pinch off on their own. Once broken

FIGURE 3-26. A cold upper-level trough followed the cold front that just passed to the southeast. Scattered, showery popcorn cumulus and even this cumulonimbus cloud boiled up in the turbulence outside my car along the busy northbound freeway.

off from the more predictable migrating winds and weather patterns, the cold upper-level cutoff low can wander somewhat independently, destabilizing the weather wherever it goes. You might recall from chapter 2 the similar disturbances that slid south into the Great Basin during autumn; I called them inside sliders. This is how we got the weather idiom "a cutoff low is a weather forecaster's woe." Cutoff lows misbehave like erratic and unpredictable sheep or cattle that have broken off from the herd, or that hyper-independent person who refuses to conform. Some of the most sophisticated forecast models struggle to predict where these little upper-level troublemakers are headed and why they don't weaken as they should. More than one weather forecaster has lost sleep while attempting to predict what these little storms might do. You can survey satellite images to spot unstable popcorn cumulus clouds out over the ocean that begin to organize and rotate around a large mass of clouds shaped like a giant comma as we look down on them, signaling cyclonic spin (known as positive vorticity advection), which encourages storm formation. This cutoff low discussion can be summarized with a rather low-confidence forecast: upwind, storms may be boiling up on your horizon, threatening the next round of showers.

Just as terrestrial animals make key migratory moves when winter comes, California gray whales (or Pacific gray whales, *Eschrichtius robustus*) have evolved to migrate with the changing seasons. During winter, they abandon their summer feeding grounds in the Arctic and head toward their calving grounds in Baja California's lagoons. In spring, they return to the Arctic to complete their round-trip migration that may total more than 10,000 miles. You can see them meandering along the rugged California coast toward Baja in December and then returning up the coast during spring. Mature gray whales can grow up to 50 feet (15 m) in length and can weigh more than 70,000 pounds (35 tons). Throughout many years, I have spotted them blowing, breaching, and splashing as they skimmed around Pt. Reyes, Pt. Dume, and off some of the other points and promontories during their annual migrations. Their populations seem quite sensitive to changes in water temperatures and food sources. (Gray whales now number above twenty thousand, but with large annual ups and downs.) These and our other elegant behemoth whale species are more examples of the awesome and charismatic marine life that has adapted to navigate California ocean currents over thousands of years, currents that in turn modify the weather and climate in profound ways.

If we follow the gray whales south toward Baja, we also follow the paths of winter storms. South of the Transverse Ranges, these storms and frontal systems, crippled after passing around the mountains, often begin to reorganize as they continue south toward San Diego County. In these cases, the patterns of clouds and diminished shower activity turn the bend around Pt. Conception and Pt. Arguello and then stream southeast. They will meet the north–south-trending Peninsular Ranges (from San Jacinto Peak well into Baja), which dominate this part of the state as if to mimic the Sierra Nevada. When deepest troughs in the jet stream drop south off the coast, winter storms can briefly bring weather conditions we might associate with northern regions of the state into Orange and San Diego Counties.

Just off the coastline, when winter's moist and marginally unstable winds clear that turn around Pt. Conception, they then approach the Channel Islands. These surface bumps of land out of the smoother ocean create friction, and long trains of clouds and showers form leeward of the islands. Such "island effect" disturbances of lee clouds are most often observed downwind of the more southern Santa Catalina and San Clemente Islands. These island-generated clouds can deliver showers in an otherwise clear to partly cloudy and often blustery environment, extending all the way to the Orange and San Diego County coastlines. When they finally reach the bumpier topography and afternoon surface heating of the mainland, their rising currents are pushed higher in the atmosphere, sometimes even developing into brief thunderstorms.

You know the routine by now: when wet storms sweep more directly off the Pacific toward San Diego and encounter the Peninsular Ranges, we have yet another wet west side, dry east side story. This time, western (windward) slopes of the mountains from Idyllwild to Palomar Mountain to Julian may catch soaking orographic rainfall at lower elevations and impressive snowfall at higher elevations all the way up to the peaks and ridges. Even rarer storms may tap into those juicy ARs that can transport trains of moisture all the way from Hawaii and beyond. This may be the case when you hear about severe flooding along rivers such as the Santa Ana, Santa Margarita, San Diego, and Tijuana. Ten inches of rainfall on mountain slopes within just a day or two have been recorded even this far south during these deluges that can deliver an entire year's average annual rainfall in less than one week.

Just a few miles farther east, air must sink inland along steeply faulted slopes, causing compressional heating and drying toward the distant low-desert floor. Stuck on the dry leeward sides, Palm Springs, Palm Desert, Borrego Springs, and the Salton Sea (some of the driest places in North America) can only look west and up toward these most efficient boundary-layer

FIGURE 3-27. The San Jacinto Mountains rise over 10,000 feet (3,000 m) above sea level, forming formidable barriers to winter storms that can't make it down to Palm Springs and other desert locales. Although exceptional Pacific storms have left snowfalls measured in feet, snow seasons this far south rarely compete with the annual white-crystal dumping grounds in the Sierra Nevada to the north. Photo by Rob O'Keefe.

water dams. (The boundary layer is the layer of air immediately above Earth's surface.) The results include sunny winter days in the 70s in the Imperial and Coachella Valleys when the rest of the state is shivering. Look and listen for flocks of honking snow geese and other bird species that have migrated and settled in these mild desert winter habitats after escaping certain frozen death far to the north. They meet at the polluted remnants of the Salton Sea, a problem child we will visit in chapter 4. The annual human migrants we call snowbirds have followed the real birds from some of the same northern latitudes, pumping local economies and celebrating the low desert's mild winters.

EVERY STORM STANDS OUT IN A CLIMATE OF CHANGE

I have tracked, sensed, and measured hundreds of these winter disturbances marching across the state during my decades of weather observations in Northern and Southern California, from our beaches to our mountains and deserts. Each storm displays its own idiosyncrasies and can surprise forecasters with unexpected twists and turns, delivering mystery, magic, and excitement to weather watching . . . and new challenges to the science and art of forecasting.

In the "old days," decades before climate change more decidedly raised its ugly head, the more reliable winters were often celebrated with a series of fleeting low-pressure troughs and high-pressure ridges drifting off the Pacific and across the state. We Californians expected periods of precipitation and storminess followed by clearing and fair weather, in repetitive cycles. In recent decades, stubborn patterns seem to have gotten locked into place, resulting in long periods of record-breaking drought and wildfires, punctuated by powerful storms and devastating floods. Ask a struggling farmer or anyone from a California community that has been burned and then drowned out. Today, we start and end the rainy season anticipating and then realizing empty reservoirs, bankrupt farmers, and water rationing one year. Then we are digging out of the snowdrifts and draining water out of our flooded farms and neighborhoods the next year. Our diverse ecosystems struggle to adapt and survive. At least it makes for exciting science.

SPRINGING AND SINGING INTO SPRING

Now that you are a weather expert, you know that when the late reggae artist Johnny Nash wrote and sang his iconic song, "I Can See Clearly Now," he was sensing high pressure pushing in behind a departing storm. Like Nash, as the last winter storm blows by, we can see clearly now that the rain is gone. We're looking straight ahead to nothing but blue skies; it's gonna be a bright, sunshiny day. As the warming sun rises higher in the sky and each day grows longer, we anticipate the promising season of growth, color, and hope for the future. But wait, it's not that simple. Unlike solar and other astronomical cycles, gradual seasonal changes in our weather aren't so reliable, especially when the spring equinox comes knocking. As winter turns toward spring, we should be teased with fewer chilling returns to winter and more frequent invasions of summer. But predicting California weather too far into the future season is a fool's game, a lesson that increasingly unusual and even unprecedented weather patterns keep teaching us.

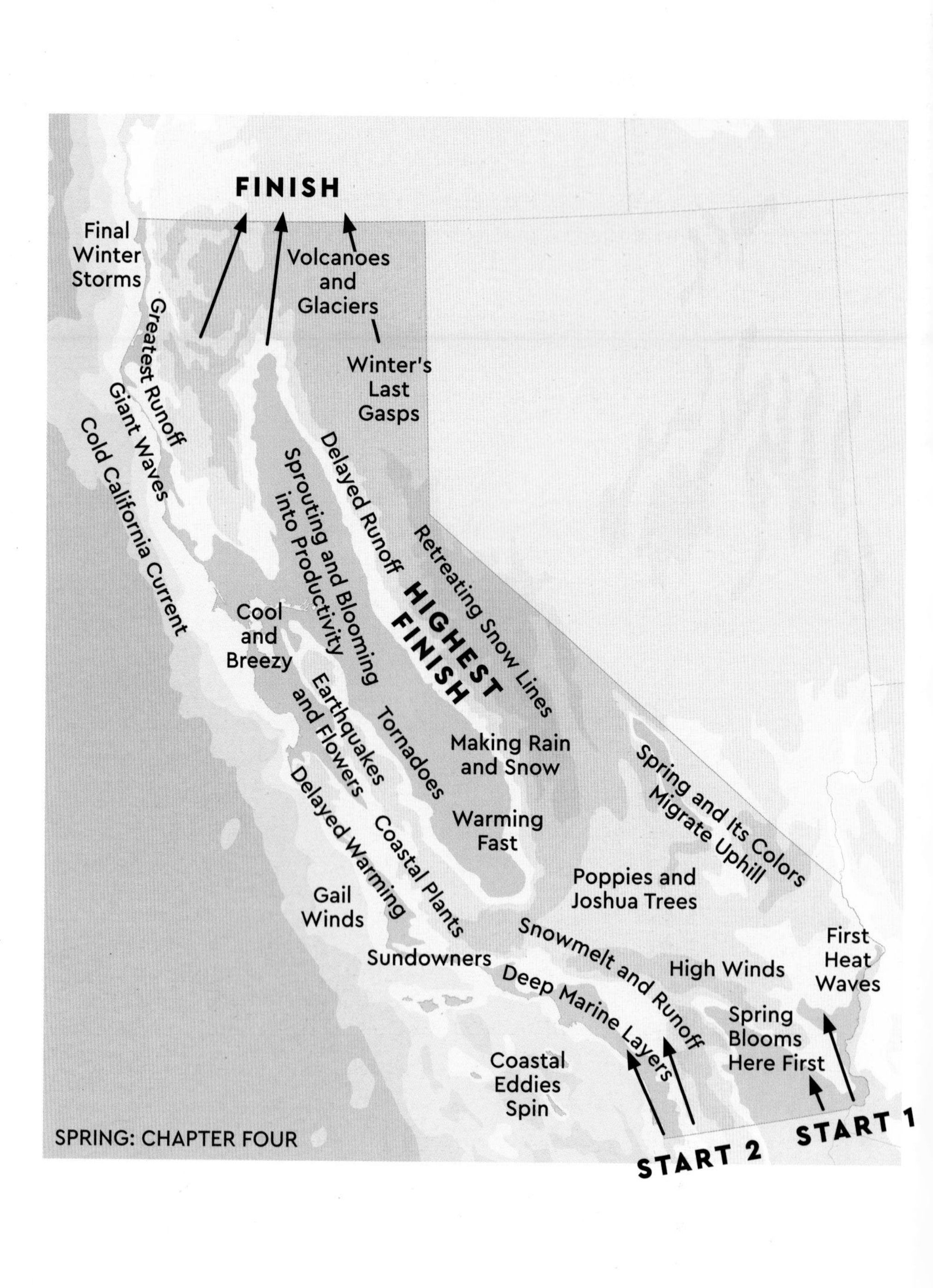

SPRING: CHAPTER FOUR

CHAPTER FOUR

SPRING'S BRIGHT REAWAKENING

There is something infinitely healing in the repeated refrains of nature—the assurance that dawn comes after night, and spring after winter.

—RACHEL CARSON, *Silent Spring*

Wake up! It's time to follow nature's coaxing that breaks us out of our hibernal cocoons and shells and into brighter, warmer spring. Just as ecosystems are rejuvenated with fresh buds, blooms, and offspring, so can we reembrace the outdoors and shed our layers for shorts and T-shirts. Hiking, spring soccer, baseball's spring training, and a host of other sandlot sports and outdoor activities call us out. It's a new season offering fresh, boundless opportunities. But before we get too confident and make too many assumptions, nature has some back-to-reality tricks up her sleeve. Winter isn't going away without a fight, and the change will take longer as we travel north or to higher elevations in the

FIGURE 4-1. Carrizo Plain National Monument, one of the finest locations to celebrate spring in California. The timing and location of the most dazzling inland-valley prairie (grassland) wildflower displays vary radically each year, especially during recent decades.

Golden State. The transition might also get ugly as surprisingly dull, inclement, and even violent weather could be the price to pay before the sunny calm season finally sets in.

Our spring-season tour of weather patterns starts where warmer weather may first appear, near the Mexican border. We will first travel north through the deserts and then up the coast and into the inland valleys. We end our weather journeys on the very highest peaks, mostly in Northern California and the Sierra Nevada.

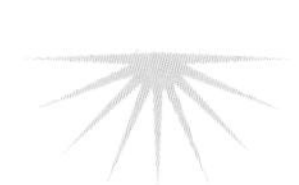

RESURGENCE BEGINS IN BAJA

When winter's constellations set in the west, Ursa Major, Bootes, Leo, and other sky-dome lights return to announce longer days with higher sun angles and the warming temperatures that follow. The warmer days and blossoming flowers that we recognize as spring first appear and then spread in southern regions of the Golden State that may not have experienced freezing temperatures, or where very brief cold snaps interrupted otherwise mild, mostly fair winter weather. You might find the first heat and flowers south of the border, down Mexico way. As the vertical rays of sun sweep back up across the equator on their way into the Northern Hemisphere, noon sun angles rapidly increase to more than 57 degrees from the southern horizon near the Mexico/California border during the vernal equinox. The noon sun is still only 48 degrees above the southern horizon near the Oregon border during the equinox, so it will take more time to loosen winter's grip on northern parts of the state. However, days will be growing longer than nights at all Northern Hemisphere latitudes as we march out of March and into the more intense sunlight and longer hours of daylight in spring.

Where soil moisture is adequate, annual wildflowers and fresh spurts of new growth quickly respond to early signs of increasing temperatures and solar radiation. As we follow nature's colorful announcements that spring is sweeping north, we recognize the start of two different spring seasons in California: the two reawakenings are separated by the highest ridges of the Peninsular Ranges, which trend south from the tallest mountain, San Jacinto Peak; through Riverside and San Diego Counties; all the way down the Baja Peninsula. Just as in California, the western cismontane (coastal) side of Baja's mountain chain showcases remarkably different landscapes compared to the eastern transmontane (inland) side. The crests of the mountains that represent the dividing line also exhibit unique higher-elevation weather patterns, climates, and plant communities.

Keep perusing your map of Baja, California. Far down the southern tip all the way to Cabo San Lucas, locals enjoy climates and ecosystems with some tropical characteristics: a summer peak in precipitation and a low-sun-season (or "winter") drought. In other words, the annual timing of their rainy season is reversed compared to that of the Golden State. Winter's polar jet streams and mid-latitude cyclones don't slump into southern Baja, or they dissipate before getting there. Also, the warm ocean waters that dominate off of western Mexico, south of Cabo San Lucas, represent a stark contrast to California's cold current that flows down our coast from the north. Because southern Baja's tropics-style rains come in the summer, its growing and blooming seasons correspond to our drought season. I have been fortunate to experience this transition—traveling from spring greening and blooming in northern Baja to the March and April drought in southern Baja.

Follow the north–south jagged line down the spine of Baja, along the crests of the Peninsular Ranges. As you imagine standing on the rocky, sparsely forested ridges of northern Baja's mountains (close to the California border), look west toward the Pacific to climates and spring seasons resembling drier versions of our state's Mediterranean coast. Weather patterns, climates, and ecosystems do not recognize political borders.

We will get back to the west side later. First, let's view east (inland) from this north–south-trending wall. We are looking toward the Sonoran Desert, which is shielded by mountains from incursions of maritime air masses off the cool California

Current. Should winter's cyclones and their frontal systems dare wander this far south, most precipitation will be blocked by these atmospheric-topographic boundary-layer dams. The rain shadow enhances a hot, arid climate that extends east across Sonora, Mexico, from the Gulf of California (Sea of Cortez). The annual monsoon spreads up from the south to deliver peak precipitation during mid-to-late summer, but winter's storms and cold snaps from the north and west are usually spent or knocked out before arriving in this inland desert. In her book *Feels Like Home,* Linda Ronstadt best describes the Sonoran Desert and its weather: "That's the desert for you—first it gives you too little, then too much, and it's ready to kill anyone who isn't paying attention."

FIGURE 4-2. Such typical spring scenes in California's low deserts stretch into Arizona, Baja, and the Sonoran Desert on well-drained slopes. Here at Anza-Borrego, budding and blooming ocotillos tower over shorter shrubs and wildflowers, such as brittlebushes, which erupt with yellow blossoms each spring.

SPRING ADVANCES INTO TRANSMONTANE CALIFORNIA

Recall how certain characteristics of the Sonoran Desert (such as some weather patterns and ecosystems) also extend north into Arizona and the low deserts of southeastern California. This region of California often experiences Sonoran-style winters, as they include an average of only two months when frost might occur and some entire winters when temperatures may barely drop to freezing in the Imperial Valley. (The coldest month in the Imperial and Coachella Valleys all the way north to Palm Springs averages daytime high temperatures just below 70°F [21°C] and overnight lows in the low 40s F [~5°C].) Add a little moisture, and spring may appear to have sprung early here as colorful wildflower displays spread across these arid valleys even before the vernal equinox. The extent and timing of color and fresh growth will largely depend on how much precipitation sneaked off of the Pacific and over the mountains in these driest desert landscapes that average just a few inches of precipitation each year.

But the blooms also depend on erratic swings in spring weather, which include late rogue winter-type storms and cold snaps, followed by early summer-like heat waves. Temperatures can plummet to near freezing behind late-season cold fronts and rebound to over 100°F (38°C) within less than a week after high pressure returns. Many years ago, an early spring heat wave greeted my students in our field class to the Sonoran Desert. When our psychrometers began measuring afternoon relative humidity near 1 percent, I worried that our handheld instruments (with their wet- and dry-bulb thermometers) were malfunctioning. Luckily, we were able to confirm our measurements at a local nature center's weather station. Such a week of extreme heat and dryness helped accelerate the annual transition from colorful desert blooms to quick wilting and seed dispersal.

Each day in early spring, especially during weekends, thousands of wildflower enthusiasts crowd Anza-Borrego Desert State Park and other low-desert hot spots, guided by social media, looking for the most riotous floral backdrop for selfies. The spectacles spring up first in these lower deserts that feel the season's early heat, and in some of the deeper desert basins to the north that include Death Valley, as winter's storm tracks and cold air masses retreat farther north. Celebrated superblooms follow the most favorable conditions, usually after wetter rainy seasons. The explosions of flowers encourage population explosions of insects (such as ants, bees, beetles, butterflies, and moths) and other pollinators. Such exceptional vernal productivity is often followed by an increase in larger herbivore and carnivore populations. More numerous and well-fed rodents, such as kangaroo rats, are leaping and darting or hiding to avoid the more numerous and well-fed rattlesnakes and other predators. The desert tortoise emerges from its long underground slumber to join the spring party. The low desert briefly comes alive with happy campers. In drier times, many of these animals may dehydrate and starve, driving their populations lower. During one of our field classes in the desert, my students were astounded by the aggressive nocturnal behavior of numerous kangaroo rats hopping into camp and by the strange behavior of coyotes that followed our caravan after we packed up and pulled out. You might have already guessed that the animals were suffering through a long, difficult drought year of famine that had followed a very wet year of plenty. These cycles repeat, some lasting for seasons; others endure for many years.

But don't blink. As the season progresses, increasingly intense solar radiation and higher temperatures quickly dehydrate the soils and the most delicate annual plants on the exposed low-desert floor, until they begin to wilt within just a few weeks. By April and May, the eye-catching waves of annual color sweep up the desert slopes to cooler, higher elevations. Low-desert cactus and desert-wash shrubs and trees, utilizing extensive root systems and water storage adaptations, attract late-spring pollinators with their brilliant flowers under a now blazing sun. Jumping cholla (*Cylindropuntia bigelovii*), also known as teddy-bear cholla, is a cactus covered with fuzzy spines that contrast with its lovely spring flowers. You will swear that it jumps at you after it seduces you to look closer or when you carelessly brush against or step on a little stem joint; it will bore its painful thorns into your shoes, clothes, or skin to get a free ride. Desert animals accidentally transport little chunks of this cactus in their fur until they can scratch them off, encouraging the propagation of this peculiarly adorable but annoying species. Cacti have adapted with water saving and storage characteristics that include those spines, which are modified leaves that cut transpiration rates and cast a little shade on the plant.

One of the more curious low-desert plants is the ocotillo (*Fouquieria splendens*), which grows spindly, thorny stems radiating out up to 10 feet high from its base. It can sprout leaves and flowers after any rain of around an inch, but it most commonly blooms during spring. The bright-red tubular flowers grow at the tips of the stems to attract hummingbirds and carpenter bees. It drops its blossoms and leaves and looks dead during long dry spells. The hardy and ubiquitous creosote bush, or greasewood (*Larrea tridentata*), so common to low- and middle-elevation deserts, also can bloom after any soaking, but is most likely to show off its brilliant yellow colors during spring. Creosote bushes grow as root-connected rings that are some of the oldest plants on Earth. One "King Clone" ring of creosote in Johnson Valley was found to be more than 11,000 years old. Although they do not tolerate long periods of extreme cold, they are successful competitors in these harsh conditions. Adaptations that include carefully timed growing and blooming, and losing leaves during the heat and drought, help them survive a range of weather extremes.

FIGURE 4-3. Various species of fleeting desert spring wildflowers blossom below larger, evenly distributed creosote bushes (not yet in bloom) in the Johnson Valley. A root-connected "King Clone" creosote ring near here indicates that this species has survived and thrived through several thousands of years of extreme weather events and climate changes.

DESERT SPRING MIGRATES NORTH AND UPHILL

As building high pressure nudges the season's final winter storms and cold air masses farther north, the spring growing and blooming season also migrates north and into higher elevations of transmontane California. A host of wildflower species decorates the understories of Joshua tree woodlands where winter snows may have dusted high-desert surfaces just weeks earlier. The Antelope Valley Poppy Reserve is flooded with pollinating insects and human visitors freeing themselves from cities and suburbs to connect with the flowers and show off California's spring poppy spectacle on social media. The intensity and timing of each display depend on a complicated seasonal dance involving soil moisture and temperatures. Across the high desert, Joshua trees (*Yucca brevifolia*) may bloom a little later in spring

FIGURE 4-4. The spring of 2023 was a banner superbloom season around the Antelope Valley Poppy Reserve.

than some of the more delicate annuals on the desert floor. Their white flowers will attract the yucca moths (*Tegeticula synthetica*) responsible for pollinating them. The presence of these Joshua trees not only signals high-desert climates between about 2,000 to 5,000 feet (600 m–1,500 m) above sea level, but each plant represents an entire ecosystem capable of supporting several animal species. There are many players in this spring orchestra, but the conductors are weather and climate.

Late spring and early summer usher the sprouting and flowering up to higher elevations where snow drifts are finally melting away with winter memories. By May and June, in the lower deserts, where the fantastical flowering first appeared, searing heat has already wilted and crumbled the helpless remains of annuals that must leave their seeds on frying-pan surfaces. As the desiccating heat and aridity expands uphill, animals seek shelter from a blazing afternoon sun by going underground or beneath scanty shade and then seeking out meager water sources between dusk and dawn. The desert returns to its high-sun, hot-season diurnal cycles: it quietly sleeps during the day and awakens at night to the dance of predator and prey hustle and bustle.

SEASONAL FITS AND STARTS POWER ENERGETIC WINDS

This gradual seasonal evolution—from winter's cold snaps and unstable transient low-pressure storms and fronts, to summer's resilient high pressure, fair weather, and stubborn heat and drought—becomes erratic. Especially in transmontane California, spring is a season of war between winter and summer weather patterns that might resemble combatants along a moving battlefront. Although summer will eventually win this war, the annual transition is punctuated by spectacular battles that can turn surprisingly violent. This is particularly evident when late-season winter-like cold fronts sweep through the state, followed by strong early season high-pressure ridges that push the fronts away in attempts to prematurely introduce summer. As a front passes, look for a canopy of clouds enshrouding the mountaintops and possibly delivering the final high-elevation snow showers of the year. After the front and low pressure race farther inland, fair-weather cumulus clouds may dot the sky, drifting with the refreshing spring breezes. A variety of lenticular (lens-shaped) cloud formations might condense in the leeward mountain waves (with names such as Palmdale and Sierra, as mentioned in chapter 3) near and behind a disturbance. But high winds will blow when strong high-pressure systems follow the season's last deep low-pressure systems. Such steep pressure gradients (extreme differences between high and low pressure over short distances) propel powerful winds that may first blow from west to east and then gradually veer more north to south as high pressure builds inland.

When this swing in direction lines up the wind with canyons and passes that concentrate and funnel the unsettled gales, it creates transmontane California's strongest windstorms. Thousands of prominent wind turbines exploit the narrow wind tunnel through San Gorgonio (Banning) Pass for just this reason. Scientists know that these prevailing winds have been blowing for thousands of years because they have transported tons of sand from where the Whitewater River flows down from the San Bernardino Mountains and into the northwestern Coachella Valley. From there, the wind has gradually pushed mountains of sand southeast across the valley. Some of this sand may even have made it all the way to the sprawling Algodones (also known as the Imperial Valley or Glamis) Dunes near Yuma, Arizona, and the lower Colorado River. (However, recent research reveals how most of the sand in that largest US dune field near Yuma was blown east from the shores of ancient Lake Cahuilla and on toward the southeastern corner of California, after being deposited by the Colorado River's ancient meandering floods.)

FIGURE 4-5. Towering wind turbines line up along the San Gorgonio (Banning) Pass wind tunnel and Interstate 10. In the background, cumulus clouds billow within rising air currents.

My students and I explored one major source of desert sand during our field classes at the Wildlands Conservancy's Whitewater Preserve. We camped where the Whitewater River spills out of the lofty San Gorgonio Wilderness to the north. This is a perfect setting to research the diverse climates, landscapes, and species that meet where the mountains are lifted dramatically above the desert along the San Andreas Fault system. It is another Golden State gem where you can stand in searing 100+° heat and look up to nearby snow, surrounded by a rich mix of plants and animals that reflect the diverse physical setting. When both a deer and a (rarer) bear wander into this desert canyon, you know you are in a transitional area between ecological regions (an ecotone) not far from the cooler and wetter forests just uphill.

Whitewater is also one of the best places to spot large populations of desert bighorn sheep (*Ovis canadensis*). This is a species that epitomizes survival under harsh conditions that include wild temperature swings and vertical cliffs. But these bighorns must drink gallons of water (which make them very sensitive to drought and other climate changes), so they may be seen hanging around water sources, which also makes them vulnerable to opportunistic predators. I have observed bighorns in other locales, such as at Joshua Tree National Park, but I have never seen so many as when our students watched them perform on the near-vertical rocky cliffs above Whitewater Canyon. As the bighorns scramble over and disturb those crumbling rock formations, smaller pieces tumble into the canyon. The deteriorating sediment can then be carried, further abraded, and deposited downhill by the river. The sand is then blown by strong winds funneling out of the pass and into dune fields, where it accumulates and migrates near the base of the towering wind turbines. These dunes stretching into the Coachella Valley are home to the endangered and legendary Coachella Valley fringe-toed lizard (*Uma inornata*). Fringed scales on their toes allow them to skim over the fine sand and even dive and disappear into it.

Human developments have interfered with what might be considered a massive sand blowing and transport machine that spills out of San Gorgonio Pass and follows Interstate 10. The winds relentlessly push the sand dunes, regardless of the houses and other infrastructure blocking nature's path. And if you get in the way of the excoriating wind that follows a spring storm, you'll quickly realize that wind-whipped sand is tougher than your skin and your car's paint job.

Still unconvinced of the power of wind and sand downwind from such gaps known for their powerful gusts? Check out the well-polished ventifacts (rocks eroded by wind-propelled sand) that have been sculpted on the hillsides facing toward the wind as it pours out of San Gorgonio and other desert passes. Meanwhile, just a few miles southwest of this wind tunnel, a perfect calm cradles some of the most cherished desert resorts in the world. Tucked at the base of those towering San Jacinto and Santa Rosa Mountains, they are shielded from most gales. Celebrities, CEOs, and other rich and famous have been attracted to the relatively tranquil narrow strip from Palm Springs to Palm Desert and the posh country club communities to the south, where residents only have to look a short distance across the valley to see the driving clouds of dust. Late-winter and early spring weather perfection in these wind shelters also helped make this valley into a Major League Baseball Spring Training mecca.

FIGURE 4-6. These ventifacts were pitted, scoured, and polished by wind-driven sand blasted out of San Gorgonio Pass and into the Coachella Valley. Similar sculpted rock outcrops may be found in isolated locations downwind of major passes throughout the state's desert landscapes.

More recent Coachella Valley development has spread into the paths of these aeolian granular migrations, as increasing populations seek out more affordable housing. (*Aeolian* means relating to or shaped by the wind.) Welcome to yet another local example of inclement affordability, as working-class Californians are squeezed toward harsher climates where they will have to endure more hostile weather (in this case, frequent wind and dust storms), but within more affordable settlements. These housing developments and a host of other engineering structures have disrupted the natural flow of sand in and around the Coachella Valley and have even threatened sand dune–dependent species such as that local fringe-toed lizard. Scientists continue to analyze the chemistry of the sand to trace its long-term sources.

FIGURE 4-7. A windstorm transports dust and sand across the Coachella Valley. Some view these turbines as a scourge on our natural landscapes; others recognize them as symbols of how we can harness clean sources of energy that would otherwise be "lost." Photo by Rob O'Keefe.

Desperate attempts to "control" nature's dust storms and sandstorms have had mixed results in the Coachella Valley. Aggressive, invasive, nonnative tamarisk (*Tamarix* sp.), also known as salt cedar, was planted in long rows parallel to the busy train tracks and Interstate 10 into the last century to serve as wind and sand breaks. But one large tamarisk tree can transpire up to 200 gallons of water per day out of precious local groundwater supplies in this sunbaked valley; alternative strategies include replanting natives (such as mesquite) that consume less water. Complexities add up as the overdrafting of groundwater has depleted local water tables until the roots of native plants can't reach them, further destabilizing the wayward dunes and reminding us how human ripple effects can become destructive anthropogenic tsunamis. You might notice dusty desert windstorms and leaping sand grains flying past local casinos and across expanding paved surfaces in the Coachella Valley. The world-renowned Coachella Music and Arts Festival has nearly been blown away more than once by April gales over 30 mph and gusts twice as strong. "Coachella Festival Goers to Battle Brutal Winds and Dust" is not the headline you want to see after investing thousands of dollars in tickets and transportation to see your favorite musicians, only to endure the aeolian spectacle that can blow in as summer drives away winter.

Travel north across the Mojave and toward Basin and Range desert basins and you will find additional isolated and sometimes impressive sand seas forming under similar conditions. Various sizes of weathered materials are stripped off of exposed rock formations by flash floods and transported and deposited downslope into desert washes and toward normally dry lakebeds. Winds whip up the finer sediments and carry them downwind of the favored canyons and passes. When prevailing

FIGURE 4-8. Desert dune native plants help stabilize parts of Kelso Dunes as afternoon cumulus clouds pop up over nearby mountains.

winds meet a mountain barrier and a calmer, more sheltered low spot, they deposit the swirls of sand that can then grow into various shapes of dunes. The tallest sand traps (Kelso Dunes) can be found in Devil's Playground south of Baker. This is near the Mojave River's final cul-de-sac, where it has deposited sediment during wetter years. Finer sands mix in the wind with weathered materials from surrounding terrains until they find their homes on these dunes. Walk up these majestic dunes during the cool season, but never try this during punishing summer days. You might hear Kelso Dunes' acclaimed singing or moaning sounds as the sand grains continually leap, strike, and settle their way along, propelled by the breeze in a rare musical version of the more common process called saltation.

FIGURE 4-9. High winds transport and then deposit weathered sand here at Mesquite Flat Dunes in Death Valley. Sinuous ridges exhibit gentler slopes on their windward sides and steeper slopes on their leeward sides. Dune shapes vary depending on wind conditions, the nature of the sand grains, and other factors.

Travel farther north into the Basin and Range to find numerous sand dunes and other aeolian features, including the accessible dune fields of Mesquite Flat near the bottom of Death Valley National Park. Similar sandscapes in Panamint Valley, Eureka Valley, Saline Valley, and south of what's left of Owens Lake are often calm in the early mornings, but can become roused after the day's sun has liberated heated thermals to swirl higher and connect to stronger upper-level winds, which then circulate toward the surface to drive sandstorms.

Throughout spring, stronger winds from the undulations of atmospheric pressure brought by passing storms carry toxic, salty dust clouds up from and over our dehydrated desert basins. Some of the briny dust was stranded, exposed, and readied for liftoff after Ice Age lakes dried up thousands of years ago. Other surfaces, such as around Mono and Owens Lakes, dried after the LA Department of Water and Power built its water diversion systems. As LA enjoys that water, it has also had to spend more than $2 billion to stabilize the toxic playa salts left behind. High-desert air quality has deteriorated to become dangerous during the windy seasons when residents would otherwise have pristine air.

Farther south, similar air quality problems have emerged on the shores of the Salton Sea in recent decades. During the twentieth century, this historic Colorado River flood faux pas became a short-lived resort and then devolved into an abandoned toxic sump and menacing regional pollution problem looking for a multibillion-dollar solution. The debacle gradually became more serious than the stench drifting with the wind during annual fish die-offs, as the Salton Sea evaporated to become more than twice as salty as the ocean. Toxic dust clouds whipped up from its dehydrated shoreline have caused a health crisis in the Imperial Valley that includes dramatically increased rates of asthma and other respiratory illnesses among the mostly working-class people (especially farmworkers) who call this place home for at least part of the year. We sometimes welcome winds that can blow away the smog in California cities, but when those gusts reach our dried-up desert basins, they are often dreaded evildoers.

FIGURE 4-10. Thousands of dead fish litter the shores of the Salton Sea. As the lake evaporates in the blistering sun and concentrates more pollutants, the wind blows salty, toxic dust into surrounding communities.

I have learned never to underestimate the power of the wind, especially in transmontane California. One spring field trip with my students to Death Valley a few decades ago stands out. The valley is aligned to funnel winds from the north, which are the ones most common from winter into spring. When we arrived at camp, the wind was so strong that some students couldn't pitch their tents, so I suggested they wait until the gusts subsided.

One ill-fated attempt ended when three of the students, utterly transfixed, could only watch as their tent launched and flew away like a kite headed into the great blue yonder. Another student's heavy pot of stew was allegedly blown right off a picnic table. In other windstorms, I have felt sand grains pelting my legs as if they were intent on stripping off my skin, and I have later plucked the sand out of my ears and scalp and from places on my body that I didn't know existed.

Driving into the remote Saline Valley in the northern Mojave Desert one late afternoon to explore the sand dunes, my friend and I settled into a breezy camp that epitomized the middle of nowhere. I noted the forecasted steep pressure gradients that would steer north winds down nearby mountain slopes, across camp, and into the valley. But as distant high pressure strengthened after midnight, a north-northeast wind had become perfectly aligned with Steel (or Steele) Canyon, which spills out just above camp. Gusts began blasting out of the canyon, so strong that they rocked our car back and forth as if it might flip over. Now shielded by the car, my tent began skidding—with me in it—until I had to get out and anchor it in place by wrestling large, heavy cobbles inside, along the sides and on each corner. As if determined to do damage, the gusts raged on, ripping holes in the tent around the rocks that just barely held it in place. Our struggles resembled scenes from a science fiction movie, as if we had landed on an inhospitably windswept planet with deadly weather flinging wind-born detritus. Nocturnal winds continued to veer around from the east until the nearby canyon was no longer aligned with the gales. The next morning dawned with a gentle breeze, allowing our exit from this mercurial desert valley on the soft, sandy dirt path that was our only road out.

The power of the wind out here goes far beyond discomfiting visiting campers. One of the most bizarre displays of the strength of desert winds is found at Racetrack Playa northwest of Death Valley, where a rock outcrop sheds its chunky weathered fragments onto the edge of the dry desert playa. Rare winter rains make the playa slick. Add the clear, freezing winter night temperatures that sometimes follow cold fronts and the mighty gales that seem to chase the storms away. The rocks find themselves imbedded in thin slabs of slippery mud and ice, and the forceful winds take over from there, pushing, skating, and gliding the rocks along the surface. You may find some of these weird "racing" rocks settled on the more recently dried playa, with their peculiar trails that tell you which direction the strongest winds were blowing as they were pushed along. Welcome to yet another otherworldly scene brought to you when and where uniquely Californian weather patterns and landscapes meet.

Especially when spring is turning toward summer and afternoon breezes become more reliable, you might find humans sailing over the dry lake beds of San Bernardino County in the Mojave Desert. As the land heats up, an intensifying thermal low pulls onshore winds across the Mojave each afternoon. Look for the "land yachts" and "dirt boats" (including custom-built rigs) sailing across perfectly flat playas that seem made for this "land sailing" that can reach thrilling speeds. It's as if the humans on the Mojave dry lakebed racetracks are trying to emulate those rocks on Racetrack Playa. When the early Chinese poet Yu Xuanji wrote in her "At Home in the Summer Mountains," "Every place the wind carries me is home," she could never have guessed that her words would fit so well in our deserts in late spring nearly 1,200 years later.

SPRING FEVER CHASES RETREATING SNOW LINES

As sun angles inch higher and each day grows longer during spring, it's useful to recall how land surfaces and inland temperatures heat up faster than the sea and coastal plains. Summer, with its stabilizing upper-level high-pressure systems, approaches and encroaches in fits and starts. Well-defined snow lines that marked winter storm snow levels begin their annual ascent, as the snow gradually and sometimes erratically melts at higher elevations. The annual melting march uphill begins earlier on Southern California slopes and becomes picturesque as folks drenched in warm sun around the pools or on the golf courses or soccer fields look up to snow-clad mountains.

Contradictions are most notable as warmth spreads north and into the Owens Valley. Spectacular scenes of sunny high-desert landscapes in the foreground clash with the well-defined Sierra Nevada snow line and frozen white carpets in the background. Such eye-catching spectacles have helped frame the Alabama Hills as one of the most filmed natural landscapes in the world. Deeper Sierra Nevada snowpacks may only slowly and stubbornly melt toward higher elevations even into late spring. As they do, the annual flood of life-giving meltwater cascades downslope, filling streams, feeding waterfalls, and occasionally pooling within inland basins that once cuddled massive lakes during the Ice Age. Cherished bodies of water (including Mono Lake, which accumulates runoff from within and around Yosemite's high country) will gather much of their annual inflow into summer, offsetting human diversions and extreme evaporation rates in the thin, dry air. The great annual melting will continue migrating north and into the Cascades, attended by dazzling wildflower displays.

As cold spells become less frequent, they represent the final interruptions in these transformations of winter to spring to summer. Winter weather patterns often say goodbye in the same way that they greeted us during autumn: in the form of those upper-level cutoff low-pressure systems we followed in chapters 2 and 3. It only takes one last cold low, slipping south from the Pacific Northwest, to remind us that winter ended just weeks earlier. The cold upper core of air encourages one last brief round of instability and lowered snow levels. More seasoned high-country locals may anticipate these exceptional perturbations, noting how they sometimes sneak through as late as mid-May. Although they can be more common and have more impact toward the north, these troublemakers sometimes encroach south into Southern California, putting a temporary freeze on spring festivals and celebrations. Such startling late-season frosty intrusions have left brief dustings of snow during May in high country around Wrightwood and Big Bear ski resorts and as far south as Idyllwild, Mt. Palomar, Julian, and other San Diego County mountains.

The most dramatic and latest-season example that I can remember of winter's refusal to yield to summer dates back decades, when I was doing field work to support my graduate research in the White Mountains. The annual roller-coaster ride of daily and weekly temperature variations throughout the eastern Sierra Nevada and Basin and Range was trending toward summer. Suddenly, one of those renegade cold upper-level low inside sliders (see chapter 2) drifted south over the region in late May during the world-famous Mule Days festivities in Bishop. This annual event around Memorial Day fills surrounding eastern Sierra Nevada and high-desert hotels, motels, campgrounds, restaurants, and other businesses. Throngs of participants and revelers can usually dress in summer attire and slather on the sunscreen to participate in mule-rodeo madness as they usher in the summer. But that year, veils of clouds and snow squalls enshrouded the surrounding mountains as temperatures plummeted to conditions more reminiscent of winter. Clouds increased in the instability until the fallstreaks of overhead snowflakes and ice crystals began extending down mountain slopes, toward the high-desert floor and its scrublands. It was as if unyielding winter had become as stubborn as the mules. Such frosty intrusions into May are rare and never last long; summer might have briefly lost a battle, but it always eventually wins the seasonal war. This is especially true in this region of California, east of the major mountain barriers, which often seems thousands of miles from the nearest ocean.

As the highest sun angles and longest days of the solstice draw near, the continent will again roast beneath dominating high pressure and the blazing sun. Spring sweeps even farther north and into the highest elevations during June in such places as the high Sierra Nevada, Mt. Lassen, Mt. Shasta, and the Medicine Lake Highlands, where roads, resorts, and campgrounds may not reopen until the snow finally melts in July. Meanwhile, another scorching summer has already set in at lower elevations; extreme evapotranspiration (all water lost from soil and plants to the air) rates begin sucking winter and spring moisture out of plants and animals that must adapt or perish. Changes can be more abrupt when strong high-pressure systems deliver record heat. Consider the daily high and low temperatures recorded at Bishop during three exceptional days, just before spring officially flipped to summer, from June 14 to June 16, 2021: 101/50, 106/52, and 107/62. More than 50°F spreads between high and low temperatures are evidence of low humidity and rarified high-desert air. By June 16, southeast flow introduced some moisture that boosted dew points, which accounts for the warmer overnight temperatures and the low of 62°F. Temperatures in Death Valley soared well into the 120s during this period, a brutal and premature start to a long, hot desert summer. Within the predictable patterns of seasonal transitions, each year tosses in surprises and curveballs, so keep watching for the annual spring awakening to find the rhythms and variations in your area.

FOLLOWING SPRING IN COASTAL CALIFORNIA

The transition from winter to summer is slower and less dramatic on the more moist slopes and valleys of western and coastal California, which face away from continental extremes and toward the Pacific Ocean. One significant factor is that the ocean takes a lot longer than the land to heat up. Water temperatures may even cool to their lowest temperatures of the year during spring, after losing energy through the winter months and then getting mixed by spring gales. In addition, the California Current and nearshore upwelling (deep, cold water rising toward the surface) are often cranking up to chill the ocean during spring. Air masses flowing off these ocean currents normally remain cool throughout the spring season. And those same moist marine air masses tend to slow and temper seasonal fluctuations, compared to the dry climates facing the continent. The difference between winter's and summer's average temperatures is just over 10°F (5.5°C) along most beaches from Oregon to the Mexican border. In some years, spring warming may hardly be noticed at the beach.

The big differences show up as the North Pacific Subtropical High pressure begins to encroach north with the sun. It slowly and inexorably pushes upper-level troughs, the Aleutian Low, and surface cyclones farther north as the season evolves. It's a reversal of fall-to-winter patterns, as if they were all taking U-turns. The water valve usually gets shut off first in extreme Southern California; the drought season will then slowly migrate north until it reaches the Oregon border by late spring. Once the North Pacific High, our great protector and drought maker, is

FIGURE 4-11. This surface weather map and satellite image shows a protective high pressure asserting itself. Just a few days before the official start of summer, the North Pacific High strengthened and expanded to dominate the eastern Pacific. *Source:* NOAA/National Weather Service data as displayed by San Francisco State University.

reestablished and anchored in place, it might occasionally wobble around and expand and contract, but it will always rule again over the state's Mediterranean late spring and summer.

Let's start in northern Baja again, but now on the coastal side of Mexico's Peninsular Ranges. The farther north you move toward California, the more in tune the Pacific-facing ecosystems are to the arrival of winter water; they often burst forth with new growth and blossoms just following the few winter storms that sprinkled life-giving water in the slightly chilled air. So, after a wet winter, the hills are already splashed with color and fresh growth when the spring equinox finally arrives. Along the coastal side of the Baja Peninsula, south from Tijuana and on past Ensenada during spring, you will see semiarid coastal and desert scrub plant communities celebrating the little bit of moisture that accumulated in the soil during the low-sun season. The color spreads north across the border, right up into coastal California, where temperatures remained relatively mild as winter storms dampened the soil. Moistened plant communities quickly respond to higher sun and longer days, even when spring temperatures may remain cool.

There are countless unique and fascinating plant and animal species that have adapted to these Mediterranean climate cycles, so we will consider just a few here. Fuchsiaflower gooseberry (*Ribes speciosum*), a quintessential California Mediterranean species, grows on coastal slopes from northern Baja to Central California. It erupts with attractive new leaves during the first rains of winter; showy red, tube-shaped flowers that conspicuously dangle from its branches will soon follow. Its flowers and berries are important food sources for a variety of birds and other animals. But it exhibits a split personality, divided by the seasons. It is drought deciduous, meaning that it will lose its flowers and leaves by summer, when it turns to what looks like a bunch of thorny, dead sticks. Look for it in shady locales. You can see how it celebrates the mild, wet winters and springs and sleeps through the drought of summer and early fall. Laurel sumac (*Malosma laurina*) is one of the more common coastal sage and chaparral species, growing from northern Baja up to San Luis Obispo. These large shrubs or small trees usually flower after the gooseberry shrubs, in late spring and early summer, and also offer vital support to wildlife. They don't lose all their leaves and, like most coastal Californians, don't tolerate extended periods of extreme cold, as they are adapted to the region's mild Mediterranean seasonal cycles. Watch their taco-shaped leaves perk up after the first rains break the annual drought. You will find a host of other plants and animals thriving in winter and through spring in these mild coastal climates. Why would any organism want to go dormant or hibernate during the rainy season in a place that never gets too cold, where life and death are determined by the where and when of water?

Deeper in the canyons and along creeks, Western (or California) sycamore (*Platanus racemosa*) is another common native to California, but it always signals the presence of water. Sycamore grow near springs and streams or at the bottom of sheltered canyons with high groundwater levels where they can keep their roots damp. These bewitching trees, with their attractive white-and-tan bark, can grow over 30 feet (9 m) high, with a trunk circumference of up to 10 feet and with enormous, relatively soft, palmate leaves up to 10 inches (25 cm) wide. They tower in clusters from Baja through well-watered coastal plains all the way up into the Sacramento Valley. When they drop limbs, large cavities are opened to house animals such as ringtails and raccoons. Sycamore also represent habitats for monarch butterflies, western tiger swallowtail butterflies, and various nesting bird species. Research suggests that sycamore were stranded along watercourses after being far more widespread during cooler and wetter Ice Age conditions. The habit of shedding leaves during the winter could be a holdover from colder times. Their year-round sources of local water within riparian habitats allow them to thrive through summer droughts.

The California fan palm (*Washingtonia filifera*) is our only native palm, but it is native only to our desert oases on the leeward side of the mountains. Their limited natural range is in low-desert canyons, often where faults create weaknesses in rock formations that allow groundwater to seep to the surface. They line up along and downstream of these natural desert springs. But they thrive as transplants and now commonly pop up as weeds within mild climates along the coast, as if they'd been there all along. Many folks are surprised to learn that all palm trees now growing on the coastal side of the Golden State's mountains had been imported or are descendants of ancestors transplanted from those desert oases. Palms are not naturally widespread—because there are no tropical climates in the Golden State—but the winters are usually warm enough (especially in Southern California) to grow tropical and subtropical plants as long as someone adds the water.

The amalgam of plant communities living on the coastal slopes and plains from Southern to Central California reflects a range of habitats and limiting factors, beginning with climate. (Limiting factors in an environment constrain the population and slow the growth of a species.) Because species of plants and animals interact with one another in particular communities, nurturing their own environments (including microclimates), these groupings help you understand what you sense when you scan a landscape. Coastal sage scrub (sometimes called soft chaparral) thrives where fog, low clouds, and cooler temperatures are persistent through the summer, but where winter storms have not encountered the orographic lift that can produce copious rainfall. Farther inland, coastal sage mixes in ecotones with chaparral species exhibiting harder, waxy leaves to cut evapotranspiration rates during the hotter summer drought that follows heavier winter rains. Thorny thickets of coastal chaparral are more dense than drier desert plant communities, but not as tall or lush as the trees in wetter forests. Many of these plants grow small, leathery (sclerophyllous) leaves that fold in on themselves so that the sun's rays don't land directly perpendicular to leaf surfaces, conserving water. Evergreen species such as ceanothus, toyon, lemonade berry, manzanita, and chamise can be found here. The song of the chaparral is the tap-tap-tapping of shy little gray wrentits (*Chamaea fasciata*), birds that flit and jump through the scrub, as if calling out for water with a rhythm resembling a golf ball bouncing on a kitchen floor. Hear them and you are likely in classic California Mediterranean climate and chaparral territory.

Prairie and grassland plant communities, a mix of flowering plants and grasses, often cover more hostile clay soils that swell with the rains and turn brick-like during drought, or they may spread across inland valleys that catch less rainfall than surrounding slopes in winter, but then bake in the summer sun. Many of the annuals in these communities will die away after spring flowering. Oak woodlands are found from near sea level and begin mixing with conifers at higher elevations or on wetter northern slopes. (Remember from chapter 2 how all of these plant communities have evolved with occasional fires.) Coastal slopes from Big Sur and north support riparian forests (which may include coast redwood) that line canyons downhill to near sea level. Such mixtures of forest species don't require the shelter of canyons in the cooler and wetter far north, especially as you approach the Oregon border. You can see how ecosystems telegraph specifics about the diversity of weather patterns and climates across the state. As you move north or toward higher elevations and it gets cooler with more precipitation, you can expect more biomass (total mass of living organisms) and forests.

Now we've surveyed climates and ecosystems on the grand scale: wetter microclimates on major ranges facing the Pacific Ocean nurture lusher plant communities, whereas drier, harsher climates on the sides of major ranges facing the continent present additional limiting factors to challenge living organisms.

FIGURE 4-12. Welcome to where coastal sage scrub meets chaparral in San Diego County. Yucca blossoms emerge above a variety of other blooming species in this spring scene. We are viewing down a moister, lusher, north-facing slope in the foreground and toward a more xeric south-facing slope in the background at Mission Trails Park.

One of the most important factors imbedded within all these grand generalizations is the local aspect of the particular mountain slope. Regardless of the region, plants and animals living on local south-facing slopes are facing into direct sunlight most of the year. Here, soils dry out faster at lower elevations, and higher-elevation snowdrifts melt earlier. A kind of positive feedback develops on these south slopes: once any snow cover has disappeared, darker surfaces absorb more sunlight and heat up faster, and drier surfaces do not get the benefits from evaporative cooling, leading to warmer microclimates. By contrast, local north-facing slopes receive less direct sunlight. Cold, highly reflective snow cover hangs on longer. The soils dry out more slowly, and their plant communities remain cooler and moister throughout the year. Look across any of California's rugged landscapes where relatively lusher and greener plant communities grow on north-facing hillsides, and more xeric communities struggle through more severe and extended soil-water deficits on adjacent south-facing slopes. These local alliances of microclimate and plant communities repeat themselves throughout the state, continually confirming the constant drumbeat: water is California's true gold.

As you traverse cismontane coastal and mountain slopes toward the inland valleys in Southern California, the shared hallucination of the region's delights might lead you to imagine encountering millions of flatlanders frolicking in their beach gear near abundant trees decorated with oranges below snow-capped mountains. The truth is, though, that most of the orange orchards have yielded to urban development, spring isn't usually the best beach weather even in Southern California, and the distant snow blankets can disappear rapidly during droughts and occasional spring heat waves. And it turns out that spring can get downright wild and crazy, even in the Golden State.

The transition from winter to summer can become more violent than California stereotypes might suggest. Just when passing storms and frontal systems are becoming less frequent and are pushed farther north, solar radiation is becoming more intense with each day's rising sun angle. When late-season storms carry cold air aloft, high above heated surfaces, the atmosphere can briefly become extremely unstable. Thunderstorms and waterspouts might churn over offshore waters, and when the storms drift inland, perhaps with passing cold fronts or cold cores of air in the upper levels, severe weather may result. This helps explain why folks in the LA Basin see more funnel clouds and experience more tornadoes than do residents in most other western flatlands of the same size. As these rare storms drift farther east, inland valleys can become tornado targets, and within the Central Valley, there are tornado "hot spots" in what has been called California's Kansas.

Research and data from the National Oceanic and Atmospheric Administration (NOAA) counted a total of 465 tornadoes in California from 1950 to 2021. These mostly relatively small twisters caused at least ninety injuries and more than $135 million in property damage. Such storms don't have the power or upper-level support that you will find in the world's notorious tornado alley across the Plains and Midwest US. There are no moist tropical air masses from the Gulf of Mexico and no bitter cold outbreaks from northern Canada to do battle over California. So Golden State tornadoes are not only oddities but also unable to compete with the legendary monsters in the middle of the country.

Depending on the potency of each storm season, California tornadoes have ranged in frequency from about one per year up to thirty per year for the entire state. The vast majority rank as the weakest, lowest-level twisters (EF0 and EF1 on the Enhanced Fujita Scale of 0–5 used across the country). In coastal and inland valleys, they are more likely to be generated by the strongest and most unstable winter and spring storms, with March being the peak month. In the deserts, they may occur during the late-summer monsoon season. It shouldn't surprise you that Golden State twisters are most likely to be spotted during the middle of the day, after some surface heating has steepened environmental lapse rates (the rate that temperatures change with altitude) and before it gets dark. Waterspouts are more common, and they have even swirled over the surfaces of a few lakes, such as Lake Tahoe.

Many years ago, I was fortunate to watch a waterspout organize just offshore as a dynamic wet cold front raced through. A definitive funnel began to dip and spin below dark cumulonimbus clouds. At the ocean surface, an impressive spray of water soon circumnavigated the base of this unlikely waterspout. I was looking down from a small hill and over the ocean, close enough to get a decent view through the rain. As it made landfall, the white circulating ocean spray turned into a rusty cloud of beach sand. Just as it looked as though this would become a damaging tornado event, the funnel lost its definition and dissipated like a frightened ghost, and the rising ring of sand collapsed above the cold land. Some call these weaker twisters "landspouts."

A few of the most turbulent storms can also deliver impressive brief hailstorms within violent downdrafts out of their icy thunderheads. In the discussion of thunderstorms in chapter 1, I noted how downbursts are enhanced as hail melts to liquid water, cooling the downdrafts and accelerating the freefall of water. These downbursts may spread out as powerful straight-line (horizontal) winds after they hit the surface. The sudden localized windstorms are sometimes misidentified as tornadoes;

but unlike their spinning counterparts, they will leave wrecking-ball debris such as trees and power poles arranged leaning and pointing in straight lines, all aligned parallel to the straightaway gusts. Impressive damage left by such heavy and quick hitters could tempt some to wonder whether the battle of the Titans had raged overhead. Other fantasizers might imagine violent sky conflicts involving Rodan, Ghidorah, or other favored flying dragons flapping their wings of death and destruction. As in the fantasy monster movies, the storms rapidly exhaust themselves or move on to carry out their demolition derbies in someone else's sky dome. The final scenes display an awkward "What just happened?" calm that settles across aftermath landscapes. Sequels are sure to play out at another time in another

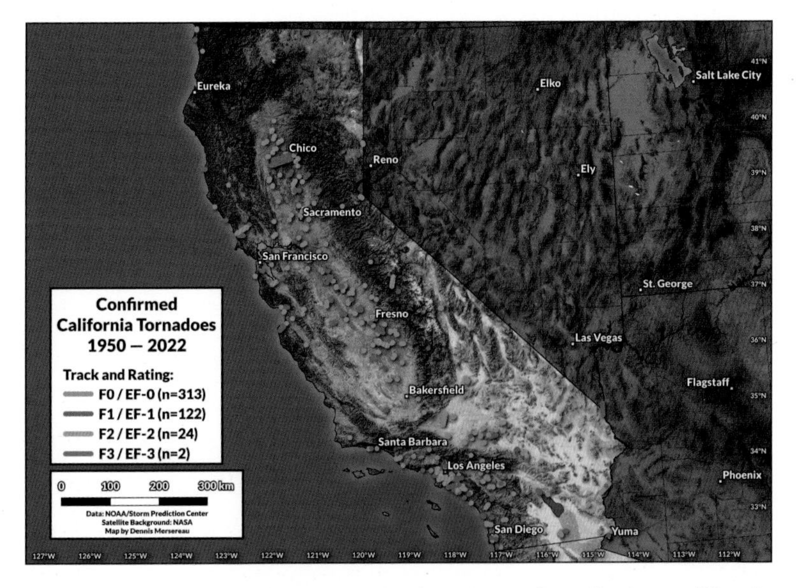

FIGURE 4-13. Enhanced Fujita Scale data is used to display California tornado distributions from 1950 to 2022. Afternoon sightings are most frequent. NOAA's National Weather Service Storm Prediction Center posts updated versions of this data. *Source:* Dennis Mersereau using NOAA/NWS Storm Prediction Center data.

neighborhood. Such spectacular atmospheric temper tantrums earn a lot of media attention, but they are only brief energetic perturbations in the more gradual and general progression of winter to spring to summer in California. Only the most fleeting, exceptional weather patterns can produce such violence along the Mediterranean coast, since the coming warm season coincides with the end of the storm season, and April showers will become less likely with each passing day. (By contrast, recall from chapter 1 the other potentially high-risk but late-summer severe weather events in the state's desert regions, generated by encroaching monsoons.)

COASTAL CURRENTS AND EDDIES CHURN THINGS UP

Eddies are always swirling all around us, even when the air seems calm. They sneak invisible and uninvited through open doors and windows, ruffling and shuffling whatever is light and loose as if a ghost were spinning through. These atmospheric whirling dervishes make your wind chimes clang and sing one minute, then rest in silence the next. They form when frictional forces cause the streaming wind to detour and swirl around obstacles. You may have observed larger eddies spiral around the corners of buildings and other obstructions, whirling with their momentary tumults of dust and debris. Or you can stand on the bank of a stream or river to examine the swirling eddies that spin due to friction between the stationary land surface and the flowing water, a boundary known as the wetted perimeter. Leaves and other debris are also caught in these eddies. Such spiraling eddies are quick to form, quick to migrate, and quick to die, and they are miniature examples of larger-scale features and weather makers so common within our atmosphere.

The numerous larger eddies that form near and over California are much easier to identify on satellite images, but we can also use some tricks to sense some of them from the ground. The largest recognizable eddies churn over the North Pacific and Gulf of Alaska and are what I earlier called mid-latitude cyclones. These spinners can grow to more than a thousand miles in diameter; they and their associated frontal systems produce most of the state's annual precipitation as they swirl east and inland along the Pacific Coast. Tropical cyclones are also large eddies; from summer into fall, they form off the coast of southern Mexico south of Baja. In the previous section, we discussed the rare, much smaller, but intense eddies that briefly hang and rotate below storms as funnel clouds and even rarer tornadoes. And don't forget about those relatively tiny whirlwinds referred to as dust devils, which can erupt from surfaces heated by intense afternoon sun. You might have heard how the butterfly effect and chaos theory have become simplified ways to express how the smallest of eddies could possibly set off chain reactions that propel larger eddies that could eventually change our weather. Bottom line: don't underestimate eddies of any size!

But there is another kind of eddy common to the Southern California coast. It is a local phenomenon that is also quick to organize and sometimes quick to die away, and it is difficult to predict. It could be considered intermediate in scale, somewhere between those tiny eddies that spin around your home and the giant eddies over the Pacific Ocean. It is called the Catalina or coastal eddy. Smaller versions of it (most without such popular names) can be observed almost any time of year along the jagged coastline from Oregon into Mexico. The formation and dissipation of these coastal eddies can have dramatic impacts on weather in the offshore waters and coastal plains, especially along the Southern California Bight. The arrival of these eddies can determine whether millions of people will bask and burn in warm sunshine or wait impatiently for the sun to show up after being nearly absent for days. The development of these coastal eddies can change an expected 80°–90° sun-splashed weekend into dull, gray, dreary days with temperatures hovering in the 60s on the coastal plain.

FIGURE 4-14. A late-season (June 16) storm approaches the Pacific Northwest, circulating clouds into far northern California. Northwest flow deflects around Pt. Conception and spins up a noticeable Catalina eddy with its center just southwest of the Channel Islands. The counterclockwise circulation pours a thick marine layer of low stratus clouds deep into Southern California coastal plains all the way down the Baja coast. *Source:* NOAA/National Weather Service.

Originally named after the island they often spin around, these Catalina eddies are most common during spring and early summer, but they can appear any time of year. When northwest surface winds turn out of the North Pacific High (yes, another giant eddy, but with winds pouring out of it!) and flow down the California coast, they are forced to veer around Pt. Conception and out to sea before entering the Southern California Bight. After briefly curving toward the south and southwest around the Channel Islands, these surface winds will gradually begin to turn back onshore and toward the coast near Catalina Island. This counterclockwise spin will first carry the air toward the east and the land, then turn it more toward the northeast and then toward the north along the coastlines of San Diego and Orange Counties. By the time these strong sea breezes spin up toward LA County, the winds might be blowing from the southeast toward the northwest, the opposite direction from where they started north of Pt. Conception, completing the eddy circulation.

As the pinwheel spins, it pushes the marine layer's low clouds and fog into coastal and inland valleys along the Southern California Bight. (These swirls have been dubbed "stratocanes" by some.) Occasionally the onshore flow is so dominant, the marine layer can thicken up to a height of 4,000 or even 5,000 feet (1,500 m), blanketing everything south and west of the mountains with monotonous gray stratus clouds and even some morning drizzle. These "mist storms" are dropped from thick, low, stable stratiform clouds that drizzle billions of tiny, annoying water droplets and sprinkle embarrassingly pathetic precipitation totals. Midday sun often evaporates shallower marine layers until the coming evening, but stronger eddies can produce cloud decks thick enough to survive the normal afternoon burn-off.

These extra-deep marine layers, with their super-thick stratus clouds, block the sun during the day and hold in longwave energy at night. This results in spring and early summer days and nights on the coastal plain that exhibit little variation in high and low temperatures. As an example, consider an afternoon high of 68°F (20°C), followed by a low of 61°F (16°C), the epitome of comfortable mildness or boring monotony, depending on whether your marine layer glass is half full or half empty. When clearing finally begins, it is sometimes in a form called "reverse clearing." The eddy might begin to break up first along the beaches, allowing the sun to burn off what's left of the marine layer muck there. But because the residual low clouds were pushed and stacked up thick against coastal slopes, clearing occurs last in the inland valleys. This is quite different from summer's more common shallower marine layers with classic late-night and early morning low clouds and fog along the coast, which most often burn off all the way to the cooler coastline in the afternoons while inland areas swelter.

Such eddy circulations might quickly break up, for many reasons. Slight changes could occur in regional wind speeds and directions around Pt. Conception; a little turbulence and mixing may be introduced, perhaps after a weak front passes through; a little offshore trend develops; or the center of the pinwheel migrates away and the eddy falls apart. Regardless, even the biggest coastal eddies usually do not have upper-level support to maintain them for too long, though a new eddy might reestablish itself on the heels of the old. (Recall from previous chapters how low-pressure troughs in the upper atmosphere encourage cooler, unsettled weather, while high-pressure ridges aloft press the upper atmosphere down to create fair weather. These upper-level fluctuations also help

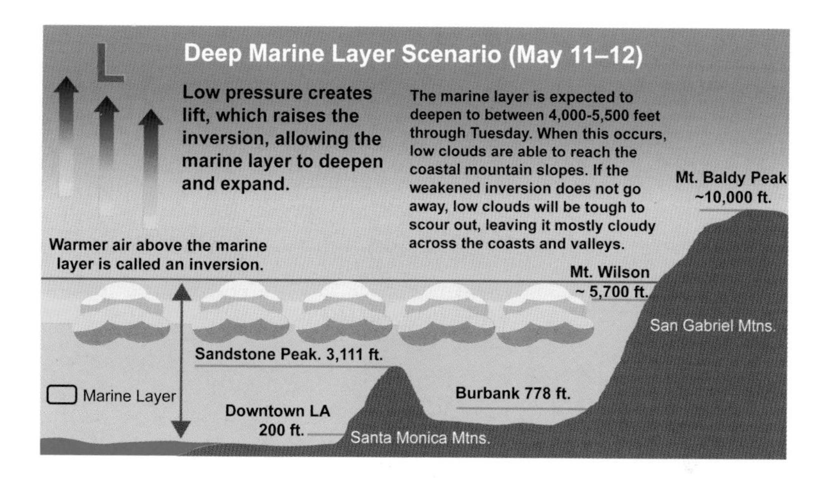

FIGURE 4-15. The deepest marine layers can produce thick, low clouds to fill coastal valleys and bank against the mountains. Such thick clouds may drop some morning drizzle and then clear very late in the day or not at all. *Source:* NOAA/National Weather Service, Los Angeles/Oxnard.

determine whether coastal eddies can expand and thicken or get squashed closer to the Earth's surface.)

Coastal Catalina eddy events are quite common during stable periods in late spring and early summer when those winds flowing northwest–southeast from the central coast are most consistent and ocean temperatures are still cool enough to keep the circulating Southern California sea breeze near its dew point. This has led to the local rhymes "May gray" and "June gloom," which gesture toward a broader truth, but disguise the still quite complex weather patterns of spring. For example, recurring coastal eddies caused marine layers to thicken from May to June 2023. San Diego was the cloudiest city in the contiguous US that May, and many locations in the Southern California coastal plain shivered through the longest string of May-June days below 80°F on record. Normally, May and June can be the cloudiest months in Southern California coastal basins if you only measure sky conditions at 8 a.m., but June afternoons are usually sunny after the diurnal "burn-off." Stubborn marine layer stratus clouds, spun up in these coastal eddies, become less common later in summer when water temperatures are warmer. They are even less common through fall and winter when there is offshore flow or when vertical turbulence mixes a more unstable atmosphere.

I've observed smaller eddies spinning all along the ragged California coast, but the Catalina eddy has created some of the most curious wind patterns. I remember, as a kid in western Santa Ana, watching dark, low clouds marching up from the south during sunset, signaling the invasion of a thicker marine layer and cooler late-spring or summer temperatures. Such occasional more energetic manifestations of stable marine layers continue to dupe visitors to California's popular coastal regions into seeking shelter from what they perceive as certain incoming summer storms. Savvy locals recognize them as harmless stratiform clouds streaming near the surface below a stable atmosphere, often spun up from an eddy. Nature puts on similar shows in the skies above most of California's coastal plains on multiple evenings throughout each spring and summer. Storm lovers are always disappointed, while tourists hoping for fair weather during their short visits are usually relieved to feel the soft feathers of summer's maritime air that might, at most, carry a little cool mist or drizzle.

I can even remember my parents noticing the low stratus clouds attacking at sunset and wondering if those menacing-looking skies were somehow connected to a fierce storm. (Mom and Dad were both transplants from climates that occasionally served out severe weather.) This led to one of those exceptional youngster epiphanies: I could actually know more than my parents about something . . . or anything. The emotional me wanted my mom to be right because I was always watching out for the anomalous severe storm that would never happen. But the rational, informed, evolving inner scientist knew better. I'd observed this little drama before, and it almost always ended with the same late-spring and summer mild-weather night and morning low-cloud chant you might remember from chapter 1. This is a moment of awareness every kid must eventually experience. Whatever your particular specialty or obsession might be, you reach a point when you finally know more about this one thing than most adults. Whether it is confusing, threatening, or liberating, this moment of empowerment founded on and fueled by knowledge is part of your life-changing journey; you will never turn back. Eventually, the same people who were once our trusted sources for answers gradually become dependent on us to answer their questions in this world of evolving role reversals. As with changing weather patterns and seasonal cycles, time waits for no one.

When the spinning motion makes it all the way up along the Ventura and Santa Barbara coastlines, you know there is an especially large counterclockwise pinwheel, an eddy dominating most of the Southern California Bight, at least near the surface. Many years ago, I was thrilled to observe an impressive textbook example of such a spinning eddy from above, looking down from over 20,000 feet (6,000 m) toward Catalina Island after rising above it all from LAX. Glad I left my window shade open.

Onshore pressure gradients increase as the North Pacific High continues to expand and strengthen through spring, propelling winds that push the cold California Current and encourage robust upwelling from deeper waters. This can keep some

FIGURE 4-16. When marine layers thicken up to about 3,000 feet or more, they are resistant to burn-off, even under the summer sun. Thick stratocumulus clouds may bank up against inland mountains and refuse to budge, leaving coastal areas to clear first by late afternoons and inland valleys under the stacks of dull gray. This is referred to as reverse clearing. Stratus clouds will likely condense and fill in again overnight.

Northern and Central California ocean temperatures depressed into the 40s F into the summer. Even my most southerly popular swimming, snorkeling, and diving spots in San Diego and Orange Counties (such as around La Jolla Cove, Crescent Bay, and Diver's Cove) require a wetsuit to guard against hypothermia during spring months. Water temperatures hovering in the 50s will shock anyone who dives in with only their skin in this game of slowly changing thermal environments. I often have to wait until later in June for my first plunge without a wetsuit, even in Southern California. And the California Current doesn't stop at the border. Depending on the year, you'll have to swim hundreds of miles south down the Baja coast to find comfortable spring water temperatures.

To observe persistent marine layers and eddies, head out into the Pacific, where coastal eddies often originate and gather momentum: the Channel Islands. Channel Islands National Park is surrounded by ocean water, so these islands have some of the mildest climates on Earth. Because ocean water temperatures remain cold into spring, and even in summer don't range above the 60s F, layers or at least wisps of stratus clouds and fog hug island slopes and spread over the uplifted, flat marine terraces all season long. The swirls that become larger coastal eddies often start around these bumps. The Channel Islands exhibit classic Mediterranean weather patterns and microclimates and the island ecosystems and landscapes to match. Average air and water temperatures around these islands vary by less than 10°F throughout the year, though greater air temperature differences are measured within microclimates toward the middle of the larger islands, especially behind topographic barriers. Rainfall is similar to that in mainland coastal regions, ranging from around 12 inches per year at some beaches, but up to 25 inches near the larger islands' highest peaks, almost all of it falling from late autumn into early spring.

On the boat ride out to the islands, you might look for dolphins and flying fish. You might even see a whale breach, especially from December through spring. More than a decade has passed since I brought groups of about thirty students to the University of California's Santa Cruz Island Field Station for a few days at a time to study the natural history and three centuries of settler impacts there. We assisted National Park Service ranger Susan Teel, who directed these science explorations as part of her research and education efforts within Mediterranean environments. During one of our adventures under the cool clouds that drift above the island's fresh green growth and colorful annual wildflowers in the spring, our students attempted to map some of the larger clusters of nonnative eucalyptus trees that had become established on Santa Cruz Island, as they crowded out native coastal sage scrub and chaparral species.

You may know that California's different species of eucalyptus were originally relocated from Australia. These trees might not be so successful here if it weren't for the Mediterranean water currents, weather patterns, and climates that help define our state. Their wood doesn't make useful building materials in the Golden State, but they've been planted to provide wind breaks, shade, and firewood, among other uses. Surveyors identified a grove near the middle of Santa Cruz Island containing what they believed to be the tallest eucalyptus ever measured. The trees must have felt at home. Naturalists and park rangers constantly debate how to deal with nonnative species in our parks and wildlands. Do we recognize eucalyptus as important cultural history features on California landscapes, or should they be eradicated for altering our ecosystems and interfering and competing with natives that are already struggling to survive in these climates? How might these invaders change local microclimates? The eucalyptus story has become more entwined with scientists' growing understanding of California weather patterns and climates than originally imagined.

Californians have discovered at least two obvious problems with eucalyptus: they shed dangerously large limbs without warning, and they act like matchsticks during fires. Residents of the Oakland Hills suffered brutally from the choice city boosters made (back in the late 1800s and early 1900s) to plant cleared hillsides with eucalyptus when the firestorm of October 1991 killed 25 people and injured about 150. That sudden wildfire terror also destroyed more than three thousand homes and quickly burned its way through more than $1 billion in property. As flames were fanned by powerful Diablo winds (a topic covered in chapter 2), numerous eucalypti seemed to morph into Roman candles across the Oakland and Berkeley Hills. What came to be known as the Oakland Hills Fire (or Tunnel Fire) spread so fast, some of the victims couldn't outrun it. Eucalyptus trees that once seemed to offer great potential to provide resources and solve problems in our Mediterranean landscapes have become notorious threats as they drop deadly limbs, disrupt natural ecosystems, and spread deadly fires.

FIGURE 4-17. Kelp beds and tide pools thrive along the Mendocino coast. These Northern California marine ecosystems depend on the California Current and robust upwelling to deliver fresh, cold, nutrient-rich water. Cool, dense, stable air masses form above the currents and then dominate the weather along the coast, especially during spring and summer.

PLUNGING NORTH INTO THE WEATHER-MAKING CALIFORNIA CURRENT

Moving farther north from the islands, it only gets colder, especially during spring. Rounding the widest part of the state at Pt. Conception and Pt. Arguello, beyond the more protected Southern California Bight, we enter increasingly turbulent Northern California waters. The coldest currents also make nearly direct hits on these exposed coastlines, and the combined local upwelling can depress water temperatures even lower during spring and into summer.

This nutrient-rich water brought up from the ocean depths supports some of the most diverse and productive marine ecosystems and fisheries in the world. Our giant kelp forests can grow up to 1 foot per day to support an astounding aggregate of biomass and diverse assemblages of animal species. Snorkeling through these marine wonderlands as they wave and oscillate with the passing swells, and swimming with garibaldi (*Hypsypops rubicundus*) and other colorful fish and playful California sea lions (*Zalophus californianus*) always rank among my most memorable nature experiences. Such seascapes are almost fantastical enough to make one forget about the water's numbing cold. If you're not up to the challenge of a teeth-chattering dive or snorkel experience, you can always explore one of California's marine aquariums. They include Steinhart Aquarium in San Francisco, Aquarium of the Pacific in Long Beach, Birch Aquarium at Scripps in San Diego, or the renowned Monterey Bay Aquarium.

The beloved southern sea otter (*Enhydra lutris nereis*) shines as one of the best examples of how marine animals have adapted to the cold waters that interact with California's Mediterranean climate. Sea otters have a double-layer coat with the densest hairs of any mammal: up to 1 million hairs per square inch (about 150,000 hairs per sq cm). By the early 1900s, these keystone-species marine mammals had been nearly hunted to extinction (only about fifty remained in California) for their dense insulating fur. After decades of protection and recovery, their populations have been fluctuating at around three thousand—less than a quarter of the state's estimated original population—from near Half Moon Bay to Pt. Conception. In such cold waters, they must eat about 25 percent of their body weight each day with a diet that includes clams, abalone, and other shelled organisms, which they break open with rocks and other tools.

Each semester, I brought students to join the traditional weekend-long marine biology field trip to the coastlines of Morro Bay and San Simeon, led by my life science colleagues. Over the decades, we observed, camped, sensed, measured, and plunged into just about every kind of weather pattern and surf condition imaginable, from calm, misty fog (usually in spring) to brilliant warm sun (usually during the fall semester) to stormy weather and high seas. Paddling a small boat around Morro Bay and watching the furry "rafts" of otters grooming themselves as they floated belly up, wrapped in buoyant kelp, is an unforgettable California nature experience. It's usually calm in the bay, but beyond the protection of the harbor, winter-like conditions extend into the spring months, with high waves and cold water.

The northern elephant seal (*Mirounga angustirostris*) is another remarkable creature that has evolved to thrive in the cold California Current. When I was in college, I enjoyed my first encounters with these marine mammals as they began hauling out at Año Nuevo State Beach. They appear awkward and bizarre on land, but are graceful and sleek and at home in

FIGURE 4-18. Wisps of coastal fog hug ancient volcanic Morro Rock. Like sea otters, peregrine falcons, gulls, and fishermen, fog banks are no strangers to Morro Rock. You can thank the cold ocean surface for the condensation that casts this typically chilling mood, especially during spring and summer.

the open water, where they spend most of their lives and can dive down to 5,000 feet (1,500 m). Many years after my first encounter, we would stop with our students to study these enormous gurgling pinnipeds on San Simeon beaches. They have adapted to these cold waters with thick layers of blubber; some males can weigh over 4,000 pounds! After these seals were hunted nearly to extinction, their protection led to yet another success story as populations have rebounded to nearly 125,000 along the California coast. After mating season, the females will arrive on the beach to give birth into March. The molting season starts in April, when you will see the peeling juveniles and females hauled out on the beach. They don't seem to mind the cold spring temperatures.

From these Central California beaches, my colleagues and I have observed with thousands of students as pelicans, cormorants, and other flocks of bird species plunged into the frigid, nutrient-rich waters, following a dense smorgasbord of schools of fish. Watch a pelican skimming just a hair's distance above the undulating waves to see how these prehistoric-looking titans have mastered an impeccable understanding of the changing small-scale pressure patterns and air currents in the atmosphere. (Pelicans were also nearly driven to extinction, first by hunting and then by DDT, but their numbers recovered after those threats were banned.)

The nutrient delivery in the upwelling encourages these various animals to adapt to living in chilly waters. We humans

FIGURE 4-19. Female and juvenile male elephant seals haul out at San Simeon Beach. The molting season is under way under typically gray May stratus clouds. You might catch the young males sparring in anticipation of future more serious and potentially dangerous fights with giant older males.

can only marvel at the inextricable links between the atmosphere, the ocean, and marine life. Our roles as both destroyers and saviors of the California coastal ecosystems tie us to these natural systems and cycles. The connections are evident everywhere in every season. For example, we were responsible for nearly wiping out the sea otter, which allowed for an overgrowth of their prey, the purple sea urchin, which then destroyed huge swaths of the Pacific coast kelp forests. Restoring the damage we've caused to the state's ecosystems continues to be a momentous responsibility.

LEARN TO FOLLOW THE WAVES

High waves arrive on California beaches in every season, including spring, when storms are still churning in the North Pacific. Breaking waves along these shorelines were generated by winds blowing across Earth's largest pool. Our beaches are extremely vulnerable to this wave energy. California's location along the Pacific Rim leaves us always facing this atmospheric and oceanic turbulence generated by some of the largest and most powerful storms on Earth. Whether nearby or many thousands of miles away, winds propagate swells that will eventually become the breakers we experience on our beaches. It all starts with friction as winds skim across the ocean surface. The largest swells or waves are generated by the strongest winds blowing over the greatest distance (known as fetch) for the longest time periods. Those three variables can combine to send monster swells to the California coast, especially from late autumn through spring.

The state's northwest-facing beaches are pummeled with waves that can climb up to 50 feet, representing some of the largest breakers ever surfed. A few of the biggest waves ever recorded in California (along the Mendocino coast) often coincided with bomb cyclones that made landfall to the north. You may never see a wave so high, but you can visit Central and Northern California after one of those magnificent storms sends its calling card thousands of miles to finally, and violently, crash on our beaches. The deafening explosions of energy combine into a symphony of roaring rhythms to warn every animal, except for the world's most skilled and seasoned surfers, to stay out of the surf zone. How does a breaker break? As swells pass over shallower water, the topography of the ocean bottom gradually interferes with their vertical rotation motions. Wave heights increase while their wavelengths decrease until the tops of the gnarly waves finally spill and curl (break) over their bases. Witness the repeated magic of refraction as waves first wrap around and exert their power on the rocky points and headlands that are exposed to the greatest wave energy; the waves will then curve into crescent shapes and disperse their energy out into more protected coves and bays until they break nearly parallel to quieter beaches, where the dissipating surf deposits sand instead of eroding it away. When storms churn in middle latitudes (such as off the coast of Asia), California's west-facing beaches may be caught in their swell windows (the direction through which swells and waves may pass). Recall from chapter 1 how summer swells from the south (including those generated by tropical cyclones) often break on our state's south-facing beaches.

Regardless of the time of year, the slope and shape of the shoreline will finally determine how waves break. If you are a strong swimmer and it is safe, jump in for a swim or ride some of these waves. Feel their power as they plunge over at angles

FIGURE 4-20. Long-period swells from distant storms combine with waves generated by local winds to wrap around rocky points and headlands and then refract into bays and coves. Stable fog and low clouds form over the cold water and then drift with spring and summer breezes here along the north coast.

and the resulting agitations and currents push you along the beach like a giant grain of sand. But never turn your back on the ocean. Prepare for the next set of monster swells that likely originated from winds and storms very far away from your beach. Atmospheric waves generating oceanic waves.

THE RESTLESS WIND TAKES CONTROL AGAIN

Once the pounding waves have carried and deposited their rivers of sand, mostly within more protected coves and bays, the wind can take over again. And as pressure patterns do battle, spring is often a very windy season at the beach. Sandy beaches can become inhospitable places of disturbance, where oscillating tides and shifting sands prevent plants from anchoring and stabilizing the coastal dunes. Gales sculpt long, linear, asymmetric ripples in the sand. Such sinuous, artistic undulations are crafted as the wind carries and pushes sand grains along, forming the gently sloped side of a ripple or dune that faces into the wind. Once they make it to the top, the sand grains will tumble down the leeward side, forming a steeper slope facing away from the wind. Although larger dunes may extend into many shapes (looking down on them), notice how the ripples often pile up parallel to one another and perpendicular to the prevailing wind. If you are caught between strong high- and low-pressure systems, admiring the repetitive beauty of these ripples and dunes may not be worth the discomfort.

Strong onshore winds have also become some of the leading characters in one of California's classic coastal battles near Pismo Beach. By 2022, Oceano Dunes State Vehicular Recreation Area was the only remaining California state park where you could drive vehicles, including non-street-legal vehicles, on the beach. Generations of families have recreated on the dunes with their off-highway vehicles that have included ATVs, dune buggies, and dirt bikes, while an adjacent fenced area protected nearby natural dune habitats. Research showed how trampling and other vehicle disturbances were destabilizing the dunes and threatening various nonhuman species. Daily vehicle traffic was also kicking up clouds of fine dust from the dunes that would be blown inland with the sea breezes and into downwind neighborhoods, creating local air-quality health hazards. To make matters worse, the strongest sea breezes tend to crank up during the middle of the day (often in spring), just when off-roaders are most active on the dunes. A struggle pitted the state parks and their recreational vehicle enthusiasts against naturalists, some local residents, and the California Coastal Commission. As of this writing, the bitter conflict has not been finally resolved, but it reminds us how Californians representing diverse special-interest groups disagree about how we should nurture, share, and use our natural resources.

While you are on the beach, you might wonder why resting birds often all seem to be facing in the same direction. Birds' wings, like airplane wings, are designed to bend and accelerate the flow of air up and over them as fast as possible. The faster air flow decreases air pressure above the wings. (This is known as Bernoulli's principle.) As the bird or plane flies faster, the difference between decreasing air pressure above the wings and relatively higher air pressure below the wings will more efficiently lift the wings and the attached bird or plane. You can see how birds facing into the wind or planes taking off into the wind will get a faster lift with less effort. So maybe those gulls you see are enjoying the sunset, but they are more likely facing into the prevailing sea breeze, which is often coming from the same direction as you look toward the sunset along the California coast. If a dog or other perceived threat appears, the birds will get an easier liftoff when launching toward the wind. And you might notice how those same birds may not appear so uniformly oriented when winds are calm or shifting.

This is also why California airports (and birds) reverse their normal air-traffic patterns when there are strong offshore winds. It makes sense to take off into the prevailing sea breezes and westerly winds until they reverse directions. So when you notice airplanes at major California airports landing and taking off toward the continent, it is likely due to offshore wind conditions. Unfortunately, as winds blow up and over an object, this decrease in air pressure may also be just enough to lift your roof off during a tornado or other powerful windstorm. Once the roof is off, such a strong wind should be able to carry it away from there.

There are many thoughtful interpretations of the late, great Maya Angelou's poem about how the free bird leaps on the back of the wind and floats downstream till the current ends. I always admired how she could have been teaching about the power of the wind to lift us as long as we are given the freedom to lift ourselves. At least we are lucky to be free to let our personal interpretations fly.

FIGURE 4-21. Sunset watching or ready for liftoff? The sun often sets into prevailing onshore winds along the coastline, so it only appears that these gulls are following the sun.

Above the frigid water, spring's saturated marine layers and drifting fog banks are more consistent and dramatic as they swirl near the points, headlands, and peninsulas. But the hour's prevailing winds will determine whether northwest-facing or southwest-facing coves, bays, and slopes are enshrouded in windward fog or basking under a giant hole zipped open within the marine gloom, protected with their topographic backs to the breeze. On a typical spring or early summer day, these sky patterns repeat themselves along the approximately 1,260 total miles (2,027 km) of the state's rugged coastline. Marine layer conditions might rule, but you will notice dramatic small-scale differences around every point and within every bay, cove, or pocket beach, atmospheric personalities that can radically change your perception of each landscape and beach day.

The Palos Verdes Hills and Peninsula exhibit some fine examples of the ways that coastal topography can disturb marine air racing toward some of the largest populations that extend across the LA Basin. Cool ocean breezes pile low clouds and fog on the windward sides of the hills. The moist air is forced to split around these obstructions that protect downwind populations from streaming fog banks. Such sea-breeze rerouting often leaves Long Beach and surroundings warmer than nearby beach communities. Along the central coast, Big Sur rarely disappoints in its ability to cast magical spells of drifting fog and mist around one turn and an open sky and sunshine around the next bend. A series of deep canyons open out into coves and sandy beaches. They are bookended by rugged resistant rock formations (the Santa Lucia Mountains) protruding out into the ocean, providing perfect settings for captivating fog dances. Northern California's Lost Coast and Cape Mendocino are other favorite (and exquisitely remote) shorelines where you can observe the walls of fog and low stratus clouds prance around the points and headlands as spring looks toward summer. Be warned that consistent late-spring and summer sea breezes along that north coast can sock in bays and headlands for days at a time until you might wonder whether the whole world has turned into a chilling foggy mist.

FIGURE 4-22. Here is where San Pedro Creek flows to the ocean (behind us) at Pacifica State Beach just south of San Francisco. On this day, local onshore flow was from the southwest (from right to left). The fog banks were struggling to get around Pt. San Pedro and other connected hilly barriers, leaving this local strand in the clear. Watch the fog attempting to cascade down leeward slopes.

Along the Santa Barbara coast, we might find one of the best examples of how coastal air mass barriers can create winds and weather unique to California. During spring, the difference between high pressure over the relatively cold ocean and lower surface pressure over rapidly heating inland regions results in strong northwest–southeast winds. These patterns increase the frequency of legendary sundowners, which are forced over or squeezed through the Santa Ynez Mountains and downward toward the narrow Santa Barbara coastal strip. Because the Santa Ynez Mountains protrude west all the way out to Pt. Conception and Pt. Arguello and trend west to east, this range represents a temporary barrier to north–south or northwest–southeast winds.

When cool northwest winds blow off the ocean and into Santa Maria, Vandenberg, and Lompoc, and encroach into the Santa Ynez Valley, they encounter the Santa Ynez Mountains obstacle. They are pushed up against the slopes and then squeezed through the passes and canyons. By the time they emerge on the south side of the range, from Goleta to the Santa Barbara coastal slopes, the dry, descending air parcels have been heated by compression: locally warm, dry offshore winds scour the coast clear. Because these winds tend to curiously peak in late afternoons and evenings, they are called sundowners. When air masses in the valleys to the north of the Santa Ynez Mountains already start out warm and dry and slopes have been heated in the afternoon sun, the additional compressional heating can push temperatures over 100°F (38°C) by the time they emerge on the south coastal side. Such occasional dry and warm spells have helped Santa Barbara earn its American Riviera moniker. If you recognize sundowners as a smaller-scale version of the Santa Ana winds we explored in chapter 2, you are right on the mark. If you noticed how they can create serious wildfire dangers and the need for red flag warnings when the slopes have been dehydrated from seasonal drought, you also nailed it. Even the spiraling horizontal and vertical eddies over the ridges and out of local canyons and other local topographic features may resemble larger-scale Santa Anas. The hot, dry bursts may be gusting out of a canyon onto one beach, while there is no wind at all just a few miles away. Some of the gusts may carry out to sea and then swirl all the way around and back onshore. The big differences between the two wind patterns are in location, scale, and wind origin and direction, and that sundowners can occur almost any time of year, but especially during spring.

I have walked these beaches when the marine layer enshrouds the entire coast all the way to Mexico. As sundowners sweep from the north and off the mountains, they suddenly push the cool moisture out and away from this narrow strip of coastline. The warm, dry air and spectacular blue skies bring welcome and startling relief from the dull gray marine layer mist. On some days, the sundowners might extend farther south through Transverse Range coastal valleys. On other days, especially from spring into early summer, the northwest winds are diverted around Pts. Arguello and Conception, and then rerouted south to gather into one of those giant eddies circulating off the Southern California Bight. California is a big state, and the weather patterns we experience in one place trigger changes elsewhere—nature connects us all.

MEDITERRANEAN PLANTS SIGNAL MILD COASTAL CLIMATES

Before we move inland and away from the domain of springtime marine layers, let's spend a little more time in our green-blanketed coastal hills. Native and nonnative wildflowers poke up from these carpets of mostly introduced grasses to celebrate the moistened soils. Grazing wildlife and cattle gorge on the spring buffet during these times of plenty. A variety of oak tree species towers above these coastal grasslands, but the archetypal coast live oak (*Quercus agrifolia*) may best define Mediterranean California. These stately trees, which can grow up to 70 feet high, thrive in dense woodlands on shadier north-facing slopes at lower elevations, but disperse or even disappear on drier south-facing slopes from Mexico to well north of the Bay Area. They grow deep, extensive root systems to gather water when it is available. Their hard, leathery leaves minimize transpiration during the drought season. Coast live oak keeps its leaves so that it is ready for the first rains; you can see them perking up and sprouting anew after the winter rains and into the spring. These trees serve as vital shelter and food sources for an astounding variety of bird and other animal species.

Pacific poison oak (*Toxicodendron diversilobum*) grows as a conspicuous shrub or vine below the coast live oak and in many other plant communities throughout cismontane California, from Mexico to Oregon. The "leaves of three" turn red during the dry season and may fall off by autumn; the remaining stems are just as toxic, causing serious skin rashes in humans. This common Golden State shrub provides food for numerous birds and other animal species. Most humans learn to keep their distance or suffer the consequences. Poison oak stems are smooth, in contrast to nearby wild berry plants that grow hairy, thorny stems.

SPRING SPREADS INTO INLAND VALLEYS

If there is a mascot tree in California's inland valleys, where you will find cool winters and hot, dry summers, it might be the valley oak (*Quercus lobata*). This is a majestic, fast-growing oak that can reach a height of 60 feet. It is common to Central and Northern California from north of Los Angeles to north of Redding. You will most often find it decorating slopes covered with rich, damp soils, and especially clustering near or lining river and stream floodplains. Unlike the coast live oak, the valley oak grows large, softer leaves and sheds those leaves in winter after turning color in autumn. Although the two species can be found mixed together, the valley oak's range signals the valley microclimates and habitats that often contrast with the milder coastal slopes of the coast live oak.

In the warmer and drier prairies (grasslands) and woodlands of the inland valleys, cold winter nights may have delayed the annual blooms. But once spring has sprung, the reawakening spreads fast as days lengthen and the sun moves toward its solstice peak. Flatlands farther inland get hot fast; afternoon temperatures may soar above 100°F (38°C) before the summer solstice. By early spring, winter's radiation (tule) fog thins and is limited to early mornings, until it evaporates into a memory for the season. Blooming seasons tend to be shorter in these inland locales more distant from sea breezes; the hills may be alive with the sound of music and abundance one week, only to be wilting in the baking sun the next. The ubiquitous state flower, the California poppy (*Eschscholzia californica*), is no stranger to these kaleidoscopic flatlands. It grows and blooms in almost every region of the state from border to border, below around 6,000 feet (1,800 m). Look for the brilliant yellow and orange flowers between February and September, as it is a perennial in milder climates and survives as an annual in harsher environments. Numerous slopes within and surrounding Southern California's Inland Empire light up with these spring blossoms. They've attracted trampling overflow crowds (such as around Walker Canyon, near Lake Elsinore) during spring's best superblooms.

Early visitors often wrote about spring flowering across California's sprawling prairies, noting their native forbs (herbs that are not grasses) and perennial bunchgrasses, which were well adapted to the local climates. As nonnative grasses were widely introduced and spread, the term *grasslands* became more popular to describe these plant communities. Today, it is recognized that these communities continue to nurture many of the original native species. You will find widespread efforts to restore native prairies, from the coast (such as at Pt. Reyes)

FIGURE 4-23. Thanks to Carrizo Plain National Monument, you can imagine many of the state's inland valley prairies (including the Central Valley) before the influx of European settlements. Peer across the San Andreas Fault, which helped build these landforms that now influence microclimates supporting such brief flower fantasylands.

to the inland valleys. You can visit the Carrizo Plain to experience the remains of California prairies, which once covered most of the Central Valley and other extensive coastal and inland valleys throughout the state. Here you will find what is left of an eye-catching polychromatic California ecosystem that was mostly plowed under or paved over long ago. You will not be disappointed, especially during warm spring days following wet winters.

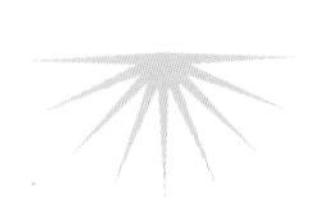

EARTHQUAKE WEATHER STRIKES AGAIN

We can only hope that Natalie Merchant wasn't a prophet when she sang about our "promised land": "San Andreas Fault moved its fingers through the ground, earth divided, plates collided, such an awful sound." But there's nothing in the lyrics about weather causing earthquakes, and for good reason. The San Andreas Fault trends alongside the sprawling flower gardens of Carrizo Plain National Monument. Rock formations along the west side of the fault have been dragged northwest for hundreds of miles over the course of millions of years, shifting from Southern toward Northern California roughly as fast as your fingernails grow. This fault system lifts mountain ranges and drops basins from well north of San Francisco to just east of the Salton Sea as it slices through the state, trending mostly northwest–southeast. You can thank the San Andreas system (and other faults and tectonic upheavals throughout the state) for these mountain ranges and valleys now acting as weather makers, as they ward off moving air masses. So in this big-picture, long-term, everything-is-connected sense, earthquakes have helped shape weather patterns and climates in California by shaping the state's topographic features.

Some Californians are convinced that earthquakes occur under certain weather conditions. For instance, I've heard Southern California residents declare that earthquakes happen during dry, warm weather. But the energy sources powering plate tectonics and earthquakes originate deep below Earth's crust, where pressures are far greater and temperatures are far hotter than in any weather experienced near the surface. Rocks that first break during earthquakes are usually many miles deep. Nevertheless, scientists do continue to research the potential influence of periodic pooling and disappearance of large water bodies above faults that are already near their breaking points. For example, recent research suggests that a shrinking and drying Salton Sea could have temporarily reduced stress on the San Andreas Fault, further delaying the next big earthquake that is long overdue there.

By contrast, the source of energy driving our weather is external. The forces and processes ruling our atmosphere start with solar energy; such energy is then cycled and distributed throughout Earth's relatively thin layer of air. Tons of dense rock materials miles below the surface are not affected by relatively insignificant changes in weather conditions above the surface. But if I had to bet, I'd guess that the next Southern California earthquake will occur during a warm, sunny spell and that the next shaker along the Northern California coast will occur on a cool, damp day. You could argue that earthquake weather is simply the weather most common to any California locale where earthquakes occur.

The energy powering the state's volcanic activity also originates deep below Earth's surface, and those larger volcanoes then act as barriers that can influence local weather patterns. Mt. Shasta, Lassen, and the Medicine Lake Highlands are excellent examples. However, Earth's greatest volcanic eruptions (such as El Chichon in 1982 and Mt. Pinatubo in 1991) are also capable of launching massive sulfuric clouds into the upper atmosphere. Such haze can then circulate around the globe and block enough solar radiation to temporarily cool Earth's atmosphere and change weather patterns. This is also a good place to note that tsunami (seismic sea waves) are generated by submarine slope failures (landslides), nearshore earthquakes, or volcanic eruptions, and are not associated with weather events.

By March and April, inland-valley cold snaps are being nudged out by incoming heat waves. Winter temperatures in the Carrizo Plain average in the mid-60s F (~18°C) by day and drop into the 30s by night, with temperatures plummeting below freezing on numerous late nights. Summer afternoon temperatures average in the high 90s (~37°C), with multiple days soaring well above 100°. But even summer nights cool nicely into the 50s (~13°C). On the inland side of the Coast Ranges, these prairies average 7–10 inches (18–25 cm) of rain each year, drier even than San Diego hundreds of miles south; whereas where the otters swim in Morro Bay, average annual rainfall is around 17 inches (43 cm), and it is considerably wetter than that on nearby west-facing slopes of the Santa Lucias. Countless earthquakes over millions of years have built these landscapes that cradle today's microclimates.

Responding to more extreme conditions in today's inland valleys, many annual spring blooms (native and nonnative) will quickly morph into seeds that wait until next winter's rains soak the soils again. But portions of the Central Valley and some other inland valleys were once covered by thick, productive strips of riparian woodlands. Hardwoods that included sycamore, willows, oaks, cottonwood, box elder, and alder supported an astounding assemblage of other plants and animals that thrived around river and stream channels through spring and summer. These tree canopies were once so extensive that they have been referred to as gallery forests. They created their own shady, cooler, and moister microclimates—until most were cleared for agriculture and other development. Much more of the valley surface you see today bakes in direct sunlight. As spring's bounty withers toward summer's drought, humans have used irrigation to defy the heat and transform many of these inland valley prairies-turned-grasslands (especially the Central Valley and Salinas Valley) into some of the most productive agricultural acreage on Earth. Today, instead of growing the resilient native plants adapted to drought, these irrigated valleys morph into humongous food factories during spring and through each summer. Such intense productivity (including irrigated farms in the searing Imperial Valley desert) continues to make California—by far—the leading agricultural state.

FIGURE 4-24. Depending on the crop, tremendous amounts of irrigation may be required to grow food in inland valleys as late spring turns toward summer. On this farm between Salinas and Monterey Bay, just enough water evaporates into the cool air from the irrigated surface to produce steam fog.

You will also find much of the Golden State's storied wine country in these inland valleys. California wines gained global fame when they started winning tasting contests many decades ago. The wide range of soil chemistries plays an important role, but grape growers and wine connoisseurs know that climate is king. Wineries exploit topography and slope aspect to orient their vines for maximum sun exposure and air circulation. Cold, wet winters followed by hot, dry summer days and cool summer nights result in the ideal sugar/acid balances that have helped make California wines supreme. The astounding diversity of wines grown in California reflects the equally astounding diversity of climates where the grapes are grown, from the north coast to the deserts.

The precise timing of cold, heat, and moisture in California's Mediterranean microclimates can determine boom or bust in wine country, and this includes unexpected weather events that might spell disaster. Cooler springs and summers delay "bud breaks" and harvests, leaving grapes vulnerable to moist periods that encourage rot. Fresh spring budding and growth thus earn plenty of anticipation and angst. Later, smoke settling from wildfires can be absorbed by the grapes, potentially causing "smoke taint" in wines. And just as climate change has been squeezing successful commercial grape growers farther north in Europe (such as from France to England), so have California growers struggled with unprecedented weather patterns that have destroyed some vintners and allowed others to experiment in regions that would have been unthinkable in decades past.

FOLLOWING SPRING WATERS INTO THE HIGH COUNTRY

As temperatures warm, spring blooms and fresh growth continue sweeping uphill in cismontane California, spreading into the Sierra Nevada and the Klamath Mountains, as higher elevations awaken from winter's freeze. Late-season storms become less frequent and dwindle into memories as the jet stream guides them far to the north.

FIGURE 4-25. The view above is across the Klamath Mountains and National Forest from near Scott Mountain as spring yields to summer at just above 5,000 feet (1,500 m) elevation.

FIGURE 4-26. Snowmelt into high-country lakes in Trinity Alps's glaciated terrain eventually runs off to nurture lower elevations in the Klamath Mountains as spring turns to summer.

RUNOFF TIMING IS EVERYTHING

During early spring in northwestern California, when storms dump more rain on already saturated soils, great swollen rivers pour out of the mountains and forests and into the ocean. The total discharge from north coast rivers and streams, which have relatively small drainage basins, averages about 28 million acre-feet of water each year, roughly 40 percent of California's annual runoff. On these forested slopes that may catch more than 100 inches (254 cm) of rain, an average of more than 40 inches (100 cm) runs off, but more than 80 inches (200 cm) can flow off the most exposed, impermeable surfaces. The raging torrents and wide water expanses of northwestern rivers during March and April—such as the Smith, Trinity, Klamath, Mad, and Eel—will gradually recede as summer approaches and each dry spell grows longer. Higher-elevation snows that once blocked roads in the Klamaths become melted memories. If you return after months of summer drought, you may not recognize the relatively meager trickles remaining in the northwest canyons.

By contrast, runoff from Sierra Nevada and Cascade high country is often delayed until later spring and summer thaws can melt the region's impressive snowpacks. This is when Yosemite's majestic waterfalls are roaring and mountain streams are flooded with icy snowmelt during warm days. This is also when hydrologists' annual snow surveys and predictions are tested. Was the state's water management infrastructure deployed correctly to emphasize flood control after a wet year or water storage and distribution after a dry year? Annual water conflicts play out among Golden State water users, pitting north versus south, private versus public, cities versus farms, and countless other diverse special interests. With 80 percent of all water used in the state still flowing to agriculture, and much of the remainder directed to landscaping, there is a cast of human characters almost forty million strong, plus millions of visitor users each year, threatening to throw or keep state water budgets in deficit. The human directors include those with the money and power, but nature will always be the producer. And it all started with winter's cyclones.

HOW TO MAKE IT RAIN OR SNOW

Cloud seeding is one of the most studied and debated forms of weather modification. It exploits natural processes to enhance precipitation from thick clouds that are potential precipitation producers. Such cloud-seeding experiments date back more than seventy years, with some "success." Remember that most of the substantial rain you have experienced in California started high in the clouds as ice and snow that eventually melted before reaching the ground. In some clouds with favorable dynamics, adding just the right number of minute particles that can act as freezing nuclei (such as silver iodide) seems to slightly enhance

FIGURE 4-27. Several feet of winter snow are finally melting in Yosemite's Mariposa Grove, nurturing the forest and feeding watercourses downstream. These giant sequoias depend on orographic precipitation that accumulates on western Sierra Nevada slopes—soon to become precious soil moisture.

FIGURE 4-28. Spring runoff is feeding Yosemite Falls and causing minor flooding in Yosemite Valley. But this is nothing compared to the flood of January 1997. That was considered the worst in Yosemite Valley history, when the Merced River crested at over 23 feet, destroying millions of dollars in infrastructure and closing the park.

precipitation totals. This happens as the freezing nuclei grow layers of ice by attracting very cold water in clouds that then freezes on to the nuclei surfaces. (This is also known as the Bergeron process.) The ice crystals grow large enough to fall through the clouds, attracting more moisture along the way and producing heavier precipitation than might have been expected. (Similar methods have been used to clear thick, cold fog banks.) This process does not work in warm clouds that may drop lighter rain and drizzle.

Rain- and snowmaking is a tricky business given that even when seeding is considered successful, a lot of uncertainty and risk management remains. We have much more to learn about weather modification, which explains why the American Meteorological Society encourages only the most well designed experimentation and research and recommends caution when people are fooling around with these natural processes. In spite of these uncertainties, as of 2022, several California agencies continued experimenting with cloud-seeding efforts, especially to enhance Sierra Nevada snowpacks. (An increasing number of studies have suggested snowfall enhancement success of up to 3–10 percent in the mountains.) Water and utility districts such as in Sacramento and the East Bay, PG&E, the Santa Ana Watershed Project Authority, and Santa Barbara and San Luis Obispo Counties continue their experimental programs, hoping for the best.

We have come a long way since Charles Hatfield roamed the drought-stricken state in the early 1900s, advancing what some believed to be his magical rainmaking skills. When he was finally hired to break the 1915 drought in San Diego, by chance a historic nearly 30 inches (76 cm) of rain fell in less than a month, causing devastating and deadly flooding. Hatfield's shenanigans even inspired a classic 1950s Western movie, *The Rainmaker*. Our weather knowledge then was barely a drop in the bucket compared to what we know today, as better-informed scientists and would-be rainmakers continue with their weather modification debates and efforts. Such struggles with nature make me suspect that the Native Americans before us better understood some of California's weather and water cycles. Consider this short excerpt recorded in 1930 by ethnographer Dorothy Demetrocopoulou. It is a small portion of an interview with a Wintu woman as it appeared in Robert Heizer and Albert B. Elsasser's *The Natural World of the California Indians*. Her oral history might summarize the Wintu's more holistic go-with-the-flow perspective toward water: "All that is water says this. 'Wherever you put me I'll be in my home. I am awfully smart. Lead me out of my springs, lead me from my rivers, but I came from the ocean and I shall go back to the ocean.'"

EXTREME EXPERIENTIAL LEARNING TURNS WET AND WILD

My learning experiences with our water cycles are generally summarized in one of the most popular quotes by Toni Morrison in her *The Site of Memory*: "All water has a perfect memory and is forever trying to get back to where it was." One summer when I was in college, I was involved in a crazy adventure that nearly turned into a deadly tragedy in the dangerous whitewaters of the Kern River in the southern Sierra Nevada. It involved a coin toss that I lost, which launched my friend on a solo ride down the rapids on our embarrassingly inadequate (and cheap) inflatable raft. After a long search downstream, I finally spotted our raft, bobbing up and down over the angry seasonal cascades, empty and upside down. Luckily, my friend eventually appeared on the opposite bank, battered and exhausted, but otherwise alive and well after a near-death experience.

Today, there are signs all along the Kern River warning visitors about its deceptive attraction. Freezing currents flowing so fast, powered by the force of gravity acting on the snowmelt, can knock you off your feet when you are just knee deep. You will occasionally hear news stories about unwary vacationers who are swept away and to their deaths along the Kern and other California rivers swollen by seasonal meltwater. One family's refreshing swim or wade can become another's terrifying, bone-breaking nightmare. You certainly shouldn't try to raft these torrents without a trained and skilled river guide. I might not be here today to tell these stories if I had "won" that coin toss so many years ago.

Nearly three decades later, one of my students, who was a river guide, convinced me to include a rafting expedition on the

American River (much farther north in the Sierra Nevada) as part of my spring semester natural history field class. She and another professional guide launched our group into the river on tested rafts that would hold up to the icy cascades. Our students experienced, firsthand, how melting snows are powered by gravity and then channeled into raging torrents carving through the mountains. After a series of alternating quiet calm and roaring bumpy sections, we eventually blasted through the narrow and thundering class IV Tunnel Chute rapid. Our experienced professional guides taught us how to keep our rafts from flipping over so that all of our students could take their hands-on hydrology learning experiences and other natural history lessons back home with them. This was a much better way to learn from our alluring but sometimes dangerous rivers and then live to tell about it. It is often true that what doesn't kill us makes us stronger . . . and wiser.

FIGURE 4-29. Although this photo doesn't depict an impressive snow or runoff year, Yosemite's Mirror Lake lived up to its name during this spring. Past floods have filled it with sediment that usually dries later in the year, resulting in a lake-less meadow during summer drought.

FOLLOW LIFE IN THE WATER

In addition to the riparian ecosystems and activities surrounding our rivers and streams, seasonal cycles support a wealth of life *in* the water. Some of the most storied species are the anadromous (migratory between fresh water and the sea) salmon and steelhead populations that call California home for parts of their lives, especially from the Sierra Nevada and Central Valley to central- and north-coast streams. The Chinook salmon (*Oncorhynchus tshawytscha*) life cycle begins in cold fresh waters, where the fish spend months to a year or so growing from juveniles, until they migrate to the ocean. They will return to the same spawning grounds to mate and die. There are four different seasonal "runs" into the Sacramento–San Joaquin River watersheds. Spring-run Chinook reenter the Sacramento–San Joaquin River drainage basins from the ocean from March into the summer. Salmon often grow to more than 40 pounds and prefer relatively deeper pools within streams.

Salmon are vital players in the state's river ecosystems for many reasons. As juveniles, they feed on insects and arthropods, and they eat fish as they grow larger. In turn, they become food sources for local wildlife, and their remains are nutrient sources (nitrogen, phosphorus, etc.) for surrounding soils and ecosystems. These salmon carry nutrients up from the cold North Pacific Ocean into forests that shade the streams and rivers, helping waterways remain shady and cold enough to support the salmon. And it all starts with those cyclones that sweep off the Pacific, delivering water evaporated from those same cold ocean currents where the salmon spend years swimming and eating. Talk about cycles! Spring-run Chinook were once the most abundant race of salmon, but they were listed as threatened in

1999 after gradually losing most of their habitats. Drought, overfishing, water diversions, warming waters, pollution caused by gold mining, dam building, various developments, and industrial activities have taken their toll. In some recent years, fishermen have chosen to forgo their catch for the year due to the collapse of the salmon population under these pressures; until more West Coast rivers can run free, cold, and deep, there won't be sufficient spawning grounds to assure the survival of future generations of salmon.

TEMPERATURE, HUMIDITY, AND EVAPORATION CONSPIRE

As we continue our expedition into the mountains, where the air is thin and temperatures extreme, we can appreciate how temperature, humidity, and wind collaborate to help determine how

FIGURE 4-30. This sunset view is toward the West Side of LA and the ocean from Griffith Park. As opposed to what you might experience in higher mountain air, you would expect the mildest days and nights within such moist coastal air masses. No problems with dehydration, hyperthermia, or hypothermia for us or our animal friends on this hike.

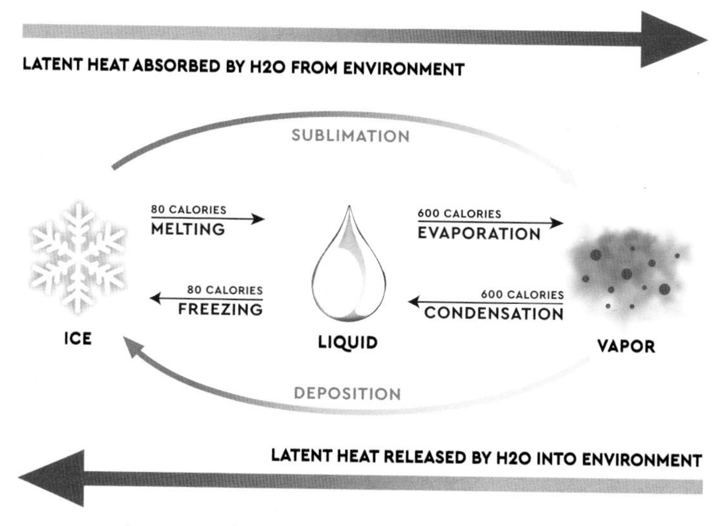

FIGURE 4-31. As water changes state, the continual exchange of latent heat has major impacts on our weather, our ecosystems, and our lives. Calories are absorbed or released between H_20 and surrounding environments during phase changes, starting with 1 gram of ice near its freezing/melting temperature.

we feel and how all living creatures survive or die. For instance, a cold wind will blow away that thin envelope of warm air that forms around your skin, causing your body to lose tremendous amounts of heat to the surrounding air. A stiff wind can make the air feel more than 20° colder than the temperature readings. Of course, if you get wet, the evaporative cooling from your skin will make you feel much colder. You could suffer from hypothermia when it is windy and you are wet, even if temperatures aren't much below 50°F (10°C). When your body has been losing heat faster than it can produce it, hypothermia will eventually set in, resulting in physical and mental collapse. Once your internal body temperature falls below 79°F (26°C), death may follow.

At the other extreme, heat exhaustion can set in after you have lost too much water (usually on hot days). Once your body temperature rises above 106°F (41°C), serious illness, fainting, and possible death will also result. It is therefore essential that you find shelter from direct sun and keep yourself hydrated when temperatures rise. All of the animals we have followed in our journeys have adapted to extreme hot and cold in their own ways, but the most efficient way to avoid overheating or freezing to death is to escape the extremes in or under a shelter. Why should we humans be so different?

Dry heat is not as oppressive as humid heat because perspiration evaporates more quickly from your skin during days with low relative humidity. Evaporative cooling accelerates further when wind is constantly sweeping away the thin layer of moist air that forms over water bodies, plants, and animals, including your perspiring skin. Every gram of water takes away around 600 calories of latent heat with it when it evaporates, leaving behind a cooler surface. So stay in the shade and drink a lot of water on hot, dry days. Even in cool mountain regions, moisture is quickly evaporated from your lungs and mouth as you breathe thin, dry mountain air, sometimes leaving you dehydrated.

By contrast, on a hot humid day, the air, already loaded with water, will not accept (evaporate) so much liquid water off of surfaces, which means that there is less cooling. You can experiment with these processes. Step into a dry breeze after your swim and feel the quick chill. Now, step out of your shower and into a steamy bathroom and notice how it takes longer to cool down. At a purely physiological level, this is why California's dry heat waves are more tolerable than the sticky, high-humidity summer steam baths found elsewhere in the United States.

We are now ready to move this discussion into our higher mountain ecosystems, where plants and animals are constantly exposed to temperature and moisture extremes. But let's first consider the risks, especially for those who (like some of my students) are afraid of animals we could be observing and studying. The media gets this wrong. It turns out that the average number of Californians who are physically attacked and killed by wild animals each year is *one or fewer*, and that includes deaths caused by great white sharks. (Dog bites, bee stings, and deer-related car accidents don't count here.) Your chances of being killed by lightning in California (thirty-four people throughout one recent 57-year period) are about the same: 0.02 out of 1 million! In contrast to the dearth of wild animal and lightning deaths, an average of about five to seven deaths are caused by flooding each year in California. Far more Californians die each year from heat exhaustion/heatstroke and hypothermia (for reasons we just covered), but the numbers are more difficult to confirm. So now that we know the real risks, let's put on our sunscreen and hat and carry our water into the mountains as spring snowpacks melt into summer.

Trees and other vegetation transpire moisture from their leaf pores; that change of state to water vapor will increase relative humidity in the shady forest and leave cool plant surfaces. By contrast, just a few feet away, the air above an exposed granite surface during a sunny afternoon will be several degrees warmer, with much lower relative humidity. Living on or near an exposed rock formation is similar to living in an urban heat island, except, with no human sources of energy, the rocks will get much colder overnight when they lose heat to the thin mountain air. Plant communities have adapted to these extremes across short distances and so have all the animals. Trees and forests provide not only shelter but milder microclimates that support a greater variety and density of species. One lightning strike and fire or other disturbance can quickly transform the

FIGURE 4-32. From the San Jacinto Mountains, across to the top of the San Bernardino Mountains, we are viewing the dividing line between more moist cismontane (to the left) and drier transmontane (to the right) Southern California. Notice a mix of plant communities from both sides of the great ranges. Bush poppies bloom in the foreground.

milder microclimate into a miniature habitat with extreme limiting factors that eliminate most species, except those who can compete in intense sunlight by day and extreme cold at night. The shelters are gone. Spring snow melts faster, and water quickly evaporates on exposed slopes as opposed to the nurturing sponge of the forest. There are many other limiting factors, but microclimates usually rule the day and night in spring and in all other seasons. We have already explored a number of species and ecosystems (including in the mountains) during our journeys across California through every season. Let's now focus on some interesting high-country critters as we climb through the melting snow and into extreme environments near the tree line. As spring gives way to summer, we will notice more animals emerging from their shelters; others are returning from their winter vacations in lower elevations or more southern locales.

California's extreme microclimates demand extreme adaptations in high-country plants and animals. At some of California's highest elevations, yellow-bellied marmots (*Marmota flaviventris*) busily graze in meadows near rock piles during high-country summers. These large furry squirrels spend much of their aboveground time basking in the sun, most likely in order to conserve energy while maintaining their body temperature. After they store about half their body weight in fat, these marmots will spend much of their lives hibernating. When they're burrowed under rocks buried in deep winter snows, their heart rates can plummet from an active 100 beats per minute to as low as 4 per minute. In hibernation, they require only six inhalations per minute, and their body temperature might decrease from an active 97°F (36°C) to near 40°F (4.5°C).

FIGURE 4-33. At such high elevations, rocky shelters and plentiful food sources make ideal habitats for yellow-bellied marmots.

By contrast, American pikas (*Ochotona princips*) remain active year-round by gathering and storing massive piles of nutritious plants in summer and foraging through the winter and spring on whatever morsels they can find in or above the snow. These little rabbit-like mammals can get all their water from their food. Although they maintain a relatively high body temperature (104°F [40°C]) to survive harsh winters, pikas do not tolerate high outside temperatures and can die when it gets hotter than 78°F (25.5°C). These charismatic critters are well adapted to alpine environments, but are not responding well to warming. Research shows that, in just the first two decades of this century, pikas' elevation range was nearly 500 feet (150 m) higher, while lower-elevation colonies have been disappearing. Biologists fear that American pikas will gradually run out of habitat as they move up their ecological island rock piles searching for cooler homes at increasingly high elevations. If you get to our high country, look for these charmers darting around and gathering sedges and grasses, drying them, and storing them. They have even been known to steal from their neighbors' poorly guarded haystacks!

At these highest elevations, melting snows deliver many benefits. They defrost slowly into the dry summer, keeping alpine soils locally saturated and delivering meltwater to riparian plant communities below. Snow cover can also protect California's few remaining glaciers from further retreat. Massive alpine (mountain/valley) glaciers that carved higher-terrain landscapes more than 11,000 years ago are long gone. Today there are still several hundred snow and ice fields remaining in California (in the Sierra Nevada, the Trinity Alps, and on Mts. Lassen and Shasta) and only about twenty named alpine glaciers that scientists can agree on (on Mt. Shasta and in the Sierra Nevada, which claims thirteen). Nearly all have been retreating during recent decades, and the rate of retreat varies each year. When heavier snow years are followed by milder summers, the ancient ice is protected under crystal shields. But recent more frequent droughts and record hot summers have been taking their toll, melting otherwise protective snow covers earlier in many summers. Then-and-now photographs of what remains reveal glaciers' alarming disintegration. As climates change, the global glacier melt contributing to sea-level rise is taking place right here in California. We are reminded how every drop of water has captivating stories to tell. "The whole moon and the entire sky are reflected in one dewdrop on the grass"—the Zen master Dōgen must have agreed.

By July 2021, snow quilts that usually cover and help insulate Shasta's glaciers had disappeared, leaving the icy masses vulnerable to accelerated melting during another summer of historic heat and drought. Ash and soot, falling out of smoke from the region's unprecedented fires, darkened surfaces and

FIGURE 4-34. Mt. Shasta's glaciers have been erratically wasting away in recent years.

hastened the melting. By late summer, meltwater carrying mud and debris was erratically cascading down the volcano. Roads, bridges, and other infrastructure were especially threatened by debris flows along Mud Creek and around the town of McCloud, including violent flows off of Konwakiton Glacier on Shasta's southern side. As climate change drives global temperatures up, at every weather event the words *unprecedented* and *historic* come tumbling over each other, like the rocks carried in the tumid mountain rivers.

ENDING OUR JOURNEY, BUT THE SHOW MUST GO ON

About one (figurative) year ago, we began our regional tours through California's seasonal weather patterns where millions of sun worshipers on the Southern California coast enjoy what sometimes seem to be endless mild summers. We finally wind up in the high country, where some would argue there are no lasting summers at all. At these elevations near 14,000 feet (4,267 m), we can't travel any higher, so it is time to end our Golden State weather and climate tours. But the weather stories don't end here, since we are all fortunate in being able to anticipate new seasons featuring nonstop meteorological mysteries and learning opportunities cycling around us. Your free ticket gets you out the door, so enjoy the sky shows. In the final chapter, we'll address some questions about weather forecasting and climate change.

CHAPTER FIVE

DEMYSTIFYING WEATHER FORECASTING AND CLIMATE CHANGE

Weather forecasting and climate change are powerfully connected. As climate change affects our lives and shapes our surroundings in profound ways, sometimes hidden and mysterious, other times astonishing and obvious, it also complicates efforts to improve weather forecasts. Our understanding of both requires knowledge of the same physical laws governing our atmosphere. And although just about everyone seems to be an expert, there are troubling misconceptions about the science behind weather forecasting and climate change. We'll clear up some of them here. When we talk about the weather, we're referring to the conditions of our atmosphere at any place for a short period of time. Climate, by contrast, refers to the combined average and extreme conditions over extended periods of time. But weather forecasters and weather models must now consider and incorporate climate changes that are impacting forecasts and forcing different outcomes compared to decades past. Whether we are forecasting the weather or understanding climate change, we are always trying to measure, understand, and anticipate the very weather elements we have explored together in this book: sunlight, energy, temperature, moisture, wind direction and speed, sky conditions, cloud patterns, precipitation. The list of variables goes on, and we have to understand how they change over space and time if we are to make sense of how they change our lives and our world.

It is amazing that forecasts spanning just one to three days have become so accurate. But as we look further into our atmospheric science crystal ball, weather elements grow foggier. By the time we extend into the next week, many variables can change in countless ways. And the butterfly effect flaps its wings of change in mysterious ways. So we paint longer-term forecasts with the broadest brush and hope for the best. Once we extend to the next season or year, we are in a no-man's land dominated by global trends that can change over months, such as the atmosphere-ocean interactions we looked at in chapter 3. All these changing and interacting variables will inevitably produce new weather patterns and climates as we extend decades into the future. So, yes, we can make some good sweeping generalizations about future climates based on the science we have learned; but if you want to know what the specific temperature or weather conditions will be at your location on a particular date in a future year, you're on the wrong planet. And now we are talking about climate change. Thus our understanding of weather and forecasting can never be separated from our understanding of climate change, but we must speak in broader generalities as we look further into the future. In this chapter, we explore how

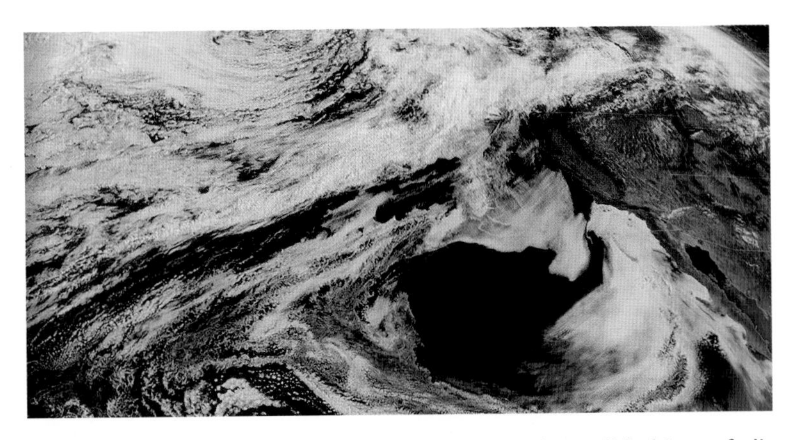

FIGURE 5-1. Atmosphere-ocean interactions and possible hints of climate change impact the forecast on the morning of June 10, 2022, as seen in this GeoColor multispectral image. Strong high pressure over California and the Southwest squeezes shallow coastal marine layers and creates dangerous heat (well over 100°F) in our inland valleys and deserts. Premature summer in California contrasts with a late-season, winter-style low-pressure system circulating toward the Pacific Northwest (top of image). *Source:* Cooperative Institute for Research in the Atmosphere (CIRA) and NOAA.

far we have progressed and where our California weather and climate journey might be heading. It will be a grand ride full of twists and turns and surprises and learning opportunities that will catapult us into the future.

AND NOW, HERE'S MY FORECAST

The best sources for weather information have drastically improved during the last century (especially for the general public) as our technologies have evolved and made Golden State weather forecasts more reliable. Let's explore the fascinating science that is used to unravel nature's weather secrets.

A BRIEF HISTORY OF WEATHER FORECASTING IN CALIFORNIA

Prior to communication technologies that became available to the public during the 1900s, weather forecasts were unreliable and awkwardly distributed. California history is peppered with countless erroneous weather forecasts that have led to historic tragedies and an appalling loss of life and property throughout the state. Misunderstood natural disasters would quickly develop into human catastrophes. We must start with a look back to see how we got here.

Starting with the Native Americans, into the Spanish and Mexican periods, and then US statehood into the early 1900s, Californians struggled to make reliable weather forecasts. Early anthropologists interviewed Native Americans and attempted

FIGURE 5-2. Layers of high and middle clouds are increasing from the west along the far northwest coast on this winter morning. We will have to consult satellite images, weather charts, and the NWS forecaster's discussions to find out if these skies and waves warn of another incoming wet storm.

to record their traditions that had been carried through generations before the European invasions. Today's Native Americans continue to save and revive these perspectives and their explanations for natural events. Scores of publications during the last several decades have attempted to summarize and interpret these life and death stories. How did the first people—who relied most on every detail of nature's ecosystem services—make weather forecasts when atmospheric changes were happening beyond their horizons? California's diverse tribes went far beyond attempting to predict the weather; they practiced many different kinds of rituals and ceremonies designed to change weather patterns in their favor. Some involved tossing water or sending up dust clouds or specific types of smoke to bring rain that might break a drought. Others incorporated sounds mimicking thunder. A few rituals, mostly in northwestern

FIGURE 5-3. Some people are still not using the wealth of forecasting sources—or their own weather observation skills—to anticipate weather changes and threats. This driver dares a fierce winter storm as it slams through the Golden Gate and into San Francisco Bay. Alcatraz can be seen in the distance.

California, were even intended to stop the storms during prolonged wet periods.

Weather conditions frequently take center stage in stories of Spanish, British, Russian, and other explorations after 1542. Success or failure (and often life and death) depended on navigating at sea and on land through thick fog or dangerous storms in mysterious places with poor maps and no weather forecasts. The first ships sailed right past the Golden Gate, at least partly because of the intense fog obscuring the entrance to the bay. After the Spanish finally settled in 1769, stories of prolonged droughts, floods, and severe weather events sprinkle the journals of missionaries and colonizers as they suffered from California's fickle weather patterns.

Ranchers, farmers, and settlers during the Mexican and early American periods struggled to anticipate and adapt to fluctuations in the state's varied climates and seasons—American

FIGURE 5-4. Mares' tails are sometimes named jellyfish clouds, for good reason. The fallstreaks are ice crystals descending from cold clouds that then sublimate (turn to vapor) and disappear into lower, drier layers. Such cloud formations have inspired many weather forecasting fables that come with far too many exceptions.

settlers who followed tales of gold and migratory trails couldn't understand why California's atmospheres differed so radically from those of their motherlands, where water resources were more abundant and reliable year round. Into the 1900s, Californians were surprised time and again by seasonal swings of weather that caught them off guard. Publications such as the annual Farmer's Almanac became popular weather forecasting crutches, though they were founded on highly unreliable guesses based mainly on climate statistics, and printed several months ahead of the forecast dates. Such uncertainties led to beliefs that rainmakers and other magicians had special powers not only to predict the weather but to change it. These sorcerers were certainly not as seasoned or respected as the Native American spiritual leaders who danced for thousands of years before them, attempting to calm the severe weather that brought suffering to the people. But they represented the desperation and helplessness and sometimes hopelessness people felt when their lives and livelihoods were destroyed by the weather.

As the 1900s advanced, direct observations came from land, ships at sea, and from pilots as air transportation developed and took off. The launching of weather instruments on balloons (radiosondes, which transmit weather measurements to ground stations) to soar high into the upper atmosphere and the anchoring of weather buoys at sea marked other major breakthroughs. These were and still are vital sources of information for Californians, since so much of our weather passes over the Pacific Ocean before reaching us. As communication technologies improved, these observations became more reliable and then available to larger audiences. Nevertheless, it's not hard to imagine the poor quality of weather forecasts disseminated from a system without satellites or speedy computers.

By the 1960s, the first weather satellites were in use, almost everyone had a radio, and most California families were adding televisions to their technology portfolios. These advancements, and a growing understanding of the physical laws governing our atmosphere, paved the way for more accurate forecasts and quick dissemination of this information to the general public. Although our understanding of the mathematical formulas that demystified such a complex atmosphere improved, our ability to test those formulas and physical laws in up-to-date weather predictions was hampered by the prolonged time it took to work them out and into a forecast. Upgraded computer technologies followed the satellite technologies to bring numerical forecasting into the limelight; these technologies improved accuracy and produced forecasts more quickly, a perfect combination to upgrade weather forecasts.

By the late 1900s, scientists were using advanced technologies to deliver more sophisticated and accurate weather forecasts not only to the professionals who depended on them for survival but into the homes of average Americans. You could still consult aviation and weather radio broadcasts for impressive details, just as

FIGURE 5-5. This weather station in the eastern Sierra Nevada is home to complex instruments designed to measure a variety of weather elements and related environmental conditions. Constant maintenance is required to keep its remote recordings accurate and accessible.

I had done to hone my skills. But the general public mostly relied on weather forecasts common to local news media on radio and TV and in newspapers. At first, these forecasts were delivered by news anchors and editors who might not have a clue. Then, local stations began competing for credibility and ratings by hiring mostly down-to-earth or relatively geeky personalities who might fit the stereotype of a trustworthy scientist wonk. The best TV weather personalities always credited their friends at the National Weather Service after they consulted the expert scientists there before going on the air.

WEATHER FORECASTING IN THE TWENTY-FIRST CENTURY

Today, armed with our knowledge of atmospheric dynamics, we can better appreciate the daily challenges faced by meteorologists at the National Weather Service. This is the original source of virtually all of our weather information and forecasts. That app or website or radio or TV personality you rely on for an accurate forecast? All their trails lead back to NOAA and the National Weather Service (NWS). We can thank the NWS for our weather satellites and imagery, original weather maps, and vital international cooperation (more than 190 World Meteorological Organization states and territories) required to gather data. The NWS also helps develop, update, and interpret the incredibly complicated mathematical formulas that drive increasingly accurate weather prediction models. At the NWS, you will find the professionals who get too little public acknowledgment for preparing and distributing daily forecasts that eventually trickle down to your preferred popular media.

So when local broadcasters exclaim, "Here's my forecast," they should really be saying, "Here's the forecast, from our National Weather Service meteorologists; I have made some changes so that it's more digestible and entertaining."

Any forecast starts with a range of possible climatic extremes for a given date. This is how sources such as the Farmer's Almanac gained popularity and were so frequently consulted before forecasts using the laws of physics were developed and refined. Predicting the future based on seasonal weather patterns of the past leads to, at best, wild guesses hinged on statistics. But there are times and places in California when and where this method will take you pretty far. For instance, it is reasonably safe to expect the repetitive "night and morning low clouds and fog, but mostly sunny afternoons" along the mild Southern California coast all through June. You can plan, with remarkable certainty, garden parties and beach weddings for those fair-weather afternoons. But up along the north coast in June, the marine layer will be more stubborn in the cooler misty air, and it won't be decent wedding season weather until September. In inland valleys, farmers can anticipate hot, dry days at that same time of year. By contrast, as the changeable autumn and winter seasons approach, daily uncertainties multiply. By October, camping trips and soccer games become more risky. As the fall season progresses toward winter, it's safer to wait for short-range forecasts before finalizing your plans. Consequently, long-term climatological forecasts using past statistics have their serious limitations.

In the medium and shorter ranges, forecasters have utilized a variety of methods with names that do a pretty good job of describing them: persistence forecasts, trend or steady-state forecasts, and the analogue method. Statistics and probabilities

FIGURE 5-6. When unstable Pacific storms are sweeping through, there is often a fine line between beneficial cloudbursts versus severe thunderstorms that can generate dangerous lightning and damaging winds, and even spin up a very rare funnel cloud. When one isolated storm, with violent up- and downdrafts, forms over a populated neighborhood (such as here in Long Beach), it can represent the difference between an accurate or botched forecast.

constantly remind us that every forecast method has its strengths and weaknesses.

While exploiting some of these traditional methods and considering the possible ranges of climatic extremes, today's numerical forecasts are powered by complex mathematical models and laws of physics that have proven to most accurately predict changes in our atmosphere; I sketched out the relationship of these forces in discussing atmospheric pressure and winds when we headed back to school in the introduction. The beauty of using math and the physical laws is that they may incorporate otherwise hidden variables and are objective, enabling us to filter out any personal or emotional biases. Now meteorologists can employ advanced computer technologies that make sense of seemingly endless variables to formulate such numerical forecasting in record time. Forecasters lean on the most reliable models as they practice a sort of ensemble forecasting, accounting for a range of uncertainties. They will tell you which of the many models they are employing to assemble the forecast. When the models are in sync, forecasts become more reliable and forecasters are more confident. But when the models are in conflict, forecasters must rely on their accumulated knowledge and skills to favor one model or the other, and this often means splitting the difference between dueling models. Because models usually fall more into sync as the forecast date approaches, shorter-term forecasts of a day or so ahead have become remarkably accurate.

FIGURE 5-7. The difference between this open view with scattered altocumulus clouds and a cool, gray overcast and drizzle under the lower stratocumulus deck is only a few miles and a few hundred vertical feet away. Given such local variations, weather forecasters are challenged to predict who will be in the warm and clear and who will be in the cold and mist. And this could be any mountain range just about anywhere in California.

But as we extend beyond three days, and especially beyond five days ahead, there are an infinite number of variables that can change and throw monkey wrenches into what was an otherwise reliable forecast.

You may have heard about the butterfly effect or chaos theory. It suggests that perfectly accurate forecasts will never be realized (especially for very small areas and longer time frames) because so many changing variables are involved. But scientists have made amazing progress, and the technology is allowing researchers to continually sharpen their accuracy down to smaller and smaller areas of focus. For instance, airline crashes have become far rarer today, thanks to improved local forecasts, radar and other technologies, and warning systems. As anomalous weather patterns and severe weather events increasingly threaten and then take lives and destroy billions of dollars in property every year, these advances are becoming particularly vital to Californians anxious about their weather. Although much larger populations are in harm's way than in the old days, the number of lives lost to weather disasters has been declining since historical debacles (such as the killer floods of the 1800s and early 1900s), thanks to increasing forecast accuracy and communications technologies.

INTERPRETING THE FORECAST

Because statistics and probabilities play such important roles in forecasting, we should pause here to clarify the meaning of that "chance of rain" in the forecast. Multiple other sources consistently get these definitions wrong, so only consult the NWS definitions. What does a 20 percent chance of rain for tomorrow really mean? Assuming that you live in that forecast area, there is a 20 percent chance that measurable precipitation (with a water equivalent of 0.01 inch or more) will fall on your home or job location or park—wherever you might find yourself within that forecast area tomorrow. It also means that there is an 80 percent chance that any one location in your forecast domain will measure no precipitation. However, especially when unsettled weather may bring scattered showers, measurable precipitation somewhere in your region might be likely on that same day, but only at limited locations that will probably not include yours. Thus you can see that a 50 percent chance of rain tomorrow is not just a guess. It almost certainly indicates that you will be under the influence of an active, unsettled weather pattern. You might assume it will rain at least somewhere in your forecast region—and that can be a big deal in California, especially in southern areas where the chance of rain remains near zero much of the year. Those probabilities of precipitation (POPs) are often described as a slight chance (20%), chance (30%–50%), and likely (60%–70%). The probability terms are left out for chances above 80 percent. In those cases you will simply see the forecast as rain or snow with such descriptors as "occasional," "intermittent," or "periods of."

Meteorologists are constantly monitoring and analyzing forecast models and using different forecasting methods, depending on what works best during certain weather patterns and trends. This requires frequently updating and adjusting forecasts.

The most efficient way to understand and interpret the forecast is to go directly to your NWS website. There you will find an expected wealth of satellite images, other colorful maps and graphics, and alerts and warnings, all of which will bring you right up to date. Click on "Forecast Discussion" to get the latest summary of observations and weather-pattern trends that have led to the forecast. These informative and educational essays are written and updated about every six hours by the professional meteorologists responsible for your regional forecast. They are fun to read and full of explanations from the very professionals who build the forecast from scratch.

Today's professional meteorologists who prepare these forecasts are located in regional forecast offices around the state and country. NOAA's NWS has reorganized after a long history of developments and evolutions that include scientific breakthroughs, technological advancements, efficiency improvements, and painful budget cutting. Today's NWS hardly resembles its parent government agency, which was named the US Weather Bureau until 1970. From north to south, California forecast offices have been consolidated over the years so that they are responsible for the following general regions:

MEDFORD: from southern Oregon into the northern inland edge of California

EUREKA: California's northwest coast

SACRAMENTO: the Sacramento Valley and surroundings

RENO: from northwestern Nevada into California's northern Sierra Nevada

SAN FRANCISCO: the north and central coasts and ranges down to San Luis Obispo County

HANFORD: the San Joaquin Valley, surrounding mountains, and Kern County deserts

LOS ANGELES: from about San Luis Obispo County south through Los Angeles County

LAS VEGAS: from northwestern Arizona and southern Nevada west through much of California's Mojave Desert

SAN DIEGO: Orange and San Diego Counties, including coasts and mountains down to the desert

PHOENIX: southwestern Arizona west into California's far southeastern deserts

If you think that this is a lot of offices for one state, remember where we are and who we are: the Golden State has, by far, the highest population in the US (nearly forty million), with the fifth-largest economy in the world, some of the busiest ports and airports (including the largest and busiest port in the US), the most diverse and productive agriculture industry in the nation, and the greatest water projects, all facing the Pacific Rim. Billions of dollars and millions of lives and businesses depend on the accuracy of weather forecasts in this state, which also boasts the most diverse range of weather patterns and climates, from our coastlines to our mountains to our deserts.

It turns out that the NWS is one of the country's most cost-efficient government agencies. It is funded for just more than $3 in taxes per person each year. Yet about one-third of the US economy is vulnerable to changes in weather and climate, and the annual costs of weather catastrophes have soared over $100 billion each year. In 2021 alone, there were twenty separate billion-dollar disasters (totaling $145 billion in damages nationally) that killed 688 Americans. Unfortunately, California has consistently led the nation in such calamities.

Throughout this book, I have explained how our atmosphere and ocean surfaces are constantly interacting and influencing one another. Both to forecast today's weather and to anticipate future climates, we must understand these powerful and complex air-sea interface connections. That is why such interrelationships got so much attention in chapter 3. And these relationships are just examples of the many connections that define our atmosphere and our natural world. Air and water temperatures, winds and ocean currents, air masses, moisture, and waves are just a few examples of nature's interrelated musical instruments, all essential players in California's weather and climate orchestra. But who is the conductor here? As the musicians play on, the sheet music begins to fade into the future, leaving each member of the orchestra to ad lib in unfamiliar territory. It's like watching a play with seasoned actors running out of script in a production that goes on forever. This leads us into the next section.

THE SCIENCE OF CLIMATE CHANGE IN CALIFORNIA

Climate change has been hovering over the first four chapters of this book. During each season and in every California region, it has become the wizard behind the curtain. It teases as a mysterious, nebulous ghost that we can ignore one day and that slams us with a sledge hammer the next. Adding disconcerting chaos to the very weather patterns we have been following here and trying to predict, climate change has set everyone on alert, causing people to pay greater attention to shifting weather patterns that can threaten health, property, and lives all over our state—and the world.

Too often polls ask whether people believe in global warming or climate change. My response is that it doesn't matter what we might "believe." Just as we know that Earth rotates on its axis once each day and revolves around the sun once each year, the laws of physics and chemistry regarding heat, pressure, temperature, and other weather elements show us how we are experiencing climate change on a global scale. But understanding the science behind climate change presents some unique challenges. Humans are running a truly monumental global experiment that has never been attempted, and no person can precisely predict the final outcome. Of course, climate change touches on fraught and important facets of our lives, from political affiliations to daily habits, forcing us to question and alter generations of accepted practice, from irrigation to oil extraction to our modes of travel and where we choose to live, in order to pave a path toward a more promising future. As many who are

FIGURE 5-8. Glacial polish, grooves, erratic boulders, and other glacial features indicate the direction this glacier was moving when it scraped the granitic basement of the high Sierra Nevada. The eroding alpine (mountain/valley) glaciers retreated at the end of the Ice Age more than 11,000 years ago.

studying the psychology of human responses to climate change have attested, common responses to this existential crisis are denial, depression, and panic, none of which help us understand the science of how and why the climate is changing and how best to overhaul the systems of modern life or to mitigate and adapt to what humanity may have wrought.

As you read this section, I encourage you to take a deep breath and decontaminate your mind; my goal is to strictly adhere to critical thinking and the scientific method to present the facts and evidence that will help you understand climate change science in California. I will stick to the facts. If we can objectively share this scientific analysis of climate change and learn its impact on every Golden State landscape and every Californian, whether plant or animal, healthy debates should follow. We must focus on how we can implement the wisest and most efficient personal, political, and economic policies . . . our most honest, constructive, and effective responses to what the science is telling us. As the old saying goes, "We cannot direct the wind, but we can adjust the sails."

WHAT ARE THE NATURAL CAUSES OF CLIMATE CHANGE?

Our atmosphere has been compared to an exceptionally well-tuned, well-oiled machine or to a living organism seeking equilibrium and adjusting to its surroundings so that it may perform most efficiently. Earth's natural systems and cycles were adjusting to dramatic climate changes long before humans arrived; fluctuations in energy inputs and outputs will continue long after we are gone. These gradual, natural, oscillating cycles that

FIGURE 5-9. Twisted and tortured by fierce winds and ice storms during winter and long droughts in summer, gnarly bristlecone pine (*Pinus longaeva*) struggle to survive in poor dolomite soils. Scientists have used downed wood, carbon-14 dating, and tree rings from living and long-dead bristlecones to trace paleoclimates and migrating subalpine timberlines back at least 8,000 years.

occur over many thousands of years include, first, changes in Earth's tilt on its axis caused by wobbling that occurs through roughly 23,000-year average cycles. Second, the angle of Earth's tilt on its axis changes by more than 2 degrees over 41,000 years. Third, Earth's elliptical orbit around the sun changes slightly in a nearly 100,000-year cycle. In addition to those cycles, there are gradual fluctuations in atmospheric chemistry. When you consider these all together, you can see how there would be plenty of long-term climate change over millions of years. You only have to examine the spectacular glacial landscapes of the Sierra Nevada and Klamath high country or the ubiquitous

Ice Age fossils from such locations as the La Brea Tar Pits or Diamond Valley Reservoir to see that California was cooler and wetter just more than 11,000 years ago and again through cycles (hot and cold, wet and dry) dated hundreds of thousands and even millions of years earlier.

Fossil evidence also shows how California species that thrive in cooler, wetter climates (such as sycamore and redwood) were far more widespread during those clammier times. In addition, millions of years of shifting tectonic plates have moved rock formations to different latitudes and climates and built massive mountain ranges that then served as barriers that impacted climates across the state. Succession summarizes how species, plant communities, and entire ecosystems are capable of gradually adjusting to all of these very long term natural agitations.

The long-term cycles are also punctuated by shorter-term events, such as when rare large objects from space crash through, explode, traumatize climate equilibriums, and rearrange life on Earth for years. The most recognized catastrophic impact occurred about 66 million years ago, marking the Cretaceous-Tertiary (more recently known as the Cretaceous-Paleogene boundary) mass extinction event. The greatest catastrophic volcanic eruptions, such as El Chichon (early 1980s) and Mt. Pinatubo (early 1990s), can also spew climate-changing volcanic debris and gases into the upper atmosphere that may cause immediate global cooling lasting for many months or years. By contrast, the awesome eruption of the Tonga volcano on the Pacific sea floor in 2022 ejected so much water vapor (a greenhouse gas) into the stratosphere that it was expected to cause temporary warming. These events result in relatively brief periods of atmospheric spitting and sputtering until the destabilizing intruders are expelled and/or new equilibriums are established. Sun-spot cycles are among the relatively less significant natural short-term climate influencers. But they have all altered and sometimes shocked the climates of the place we now call California.

Our recent understanding of ENSO (El Niño-Southern Oscillation) and PDO (Pacific Decadal Oscillation) cycles, which we examined in chapter 3, remind us that large-scale changes in ocean currents off the California coast can also drive dramatic short-term shifts in the state's weather and climate. As scientists gathered more reliable evidence toward the end of the 1900s, it became apparent that the combined long-term natural oscillations alone should have been driving Earth's atmosphere toward a cooler period. But just the opposite happened.

FIGURE 5-10. Transportation continues to contribute more CO_2 and other forms of air pollution than any other human activity in the state. The low clouds you see here along Interstate 405 in LA's South Bay are forming as warm, moist exhaust is emitted from the refineries and then quickly condenses around abundant condensation nuclei in the cool air—another unintended science experiment showing how car-culture humans can change the weather.

WHAT ARE THE MAIN CAUSES OF CLIMATE CHANGE TODAY?

By the end of the last century, meteorologists, climatologists, physicists, oceanographers, and chemists were measuring troubling changes in Earth's atmosphere that were driven by human activity. Scientists have long understood that our atmosphere is not primarily heated from the top down by shortwave radiation directly emitted from the sun, but mostly from the bottom up by longwave radiation emitted by land and ocean surfaces warmed by that initial sunlight; many of the different weather patterns we see in California are due to the different ways land and water surfaces warm the air above them. We all know this as the greenhouse effect. Greenhouse gases (GHGs), such as water vapor, carbon dioxide (CO_2), methane, and nitrogen oxides, allow relatively shortwave sunlight in but then selectively absorb longwave infrared radiation (heat) from Earth's surfaces. GHGs are major temperature regulators working like natural thermostats. Without them, most of Earth would freeze over, and life as we know it would not exist. The Golden State would quickly become the Frozen State.

The problem we're facing now is that the concentrations of those insulating gases are increasing fast. It's as if we are rolling up the windows of our car on a sunny day. The shortwave heat from the sun enters through our transparent atmospheric windows to heat the inside, but the longwave heat energy can't get out. There is currently far more CO_2 in our atmosphere than during the last more than one million years. Improved tools and advanced technologies are helping us extend more reliable measurements even further back into the geologic record. And we have measured and confirmed that the main source of today's increasing CO_2, methane, nitrogen oxides, and other GHGs is human activity that includes our burning of fossil fuels. Global CO_2 concentrations have increased (at accelerating rates) by about 50 percent due to these human activities since 1750, when the industrial era began. As humans dramatically change the chemistry of the atmosphere and more heat is trapped, climates are shifting relatively quickly; the laws of physics tell us that this additional energy will be distributed and reckoned with one way or another. Let's explore further.

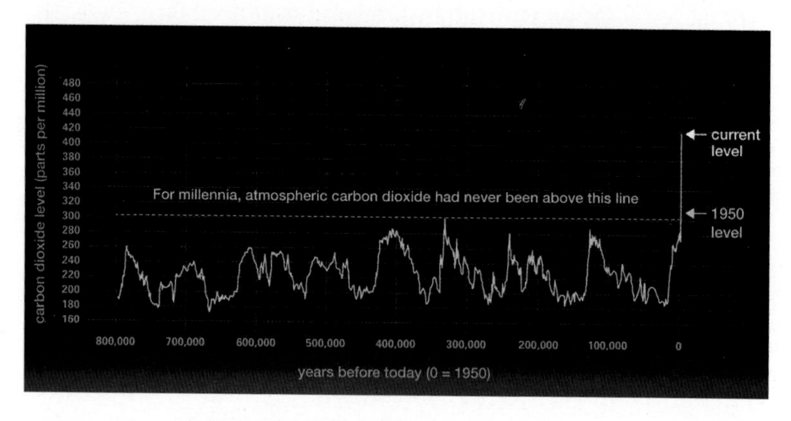

FIGURE 5-11. Global CO_2 content reconstructions from ice cores enable accurate estimates back to 800,000 years ago. Various other measurements and technologies are pushing paleoclimate estimates much deeper into the geologic record. "Current" on this curve is up to 2022. *Source:* NASA Global Climate Change and NOAA.

HOW MIGHT WE ANTICIPATE THE FUTURE?

Our climate is experiencing an unprecedented human-driven phase of relatively sudden and rapid change. Should GHG accumulations and other human interferences finally stabilize, warmer climates might also stabilize, but with very different weather patterns. However, should temperatures continue to warm above certain thresholds, positive feedbacks (when two or more changes combine to amplify initial changes and impacts) could trigger unpredictable cycles of climate chaos, with wild swings in weather patterns until new equilibriums are finally established decades or centuries later. It is impossible to perfectly predict when those thresholds might be crossed and what exact outcomes might result, because this is, to repeat my analogy, an experiment never run before on a scale never seen before. You might have heard about the Paris Agreement. One objective was to limit global warming to 1.5°C above preindustrial norms. But, as we learn more, we know that that objective is a moving target, and some scientists, as well as those impacted by climate change, are concerned that we've already passed key thresholds. This leaves scientists to

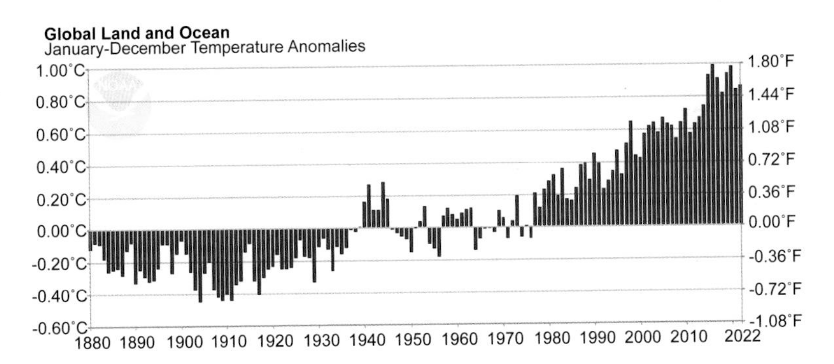

FIGURE 5-12. This chart illustrates global warming trends as impacted by ENSO cycles. During most La Niña events, global warming slows, only to accelerate again during El Niño periods. Although temperatures in California have warmed at about the same rate as they have in the rest of the world, correlation with the ENSO cycles is not so evident, perhaps because strong El Niño years tend to be stormier here, and La Niña years are often sunnier and drier. *Source:* NOAA National Centers for Environmental Information.

rely on models that use the physical laws of our atmosphere to anticipate possible scenarios, depending on how quickly atmospheric chemistry changes and how weather patterns and climate systems react to those changes. All models are showing accelerated average warming, with some local exceptions, and continued consequential disruptions to weather patterns. California is no exception.

The term *attribution science* is used by climate researchers charged with developing improved models that help us understand which dramatic changes are caused by humans. Today's range of climate model simulations suggest that if we begin cutting GHG emissions now or develop methods to sequester those emissions, global warming could eventually stall at about 2°F (1.1°C) higher than today. However, if we continue emitting as we are today, global temperatures could climb nearly 10°F (5.5°C) by the end of this century, creating an unrecognizable climate and planet. These estimated increases will be piled on top of the warming that we have already measured.

HOW IS CLIMATE CHANGE AFFECTING CALIFORNIA?

In California, as in much of the rest of the world (from pole to pole), the atmosphere warmed by an average of more than 1.5°F (about 1°C) during the last century or so. However, the rates of warming have dramatically increased since 1975, and they are different everywhere. For instance, arctic regions are warming much faster, while a few locations have measured little or no warming. The many impacts include more frequent and dramatic weather anomalies that are challenging natural and human systems to adjust and/or suffer severe and possibly permanent damage.

Our knowledge of atmosphere-ocean interactions would suggest that while average California temperatures warm, extreme temperature events that become more common in inland areas would have less impact along the coast. Scientists are studying the particulars of how the modifying influences of marine air masses off the cool ocean current might soften the climate-change blow in coastal cities, at least when it comes to temperatures. Meanwhile, tidal gauges and satellite imagery measure how sea levels have already increased by an average of about 6–8 inches (18 cm) along the California coast as global ice melts and warmer seas expand. Our state's sea-level rise is also similar to global trends during the last century or so, and it is already causing a multitude of problems as seawater encroaches farther inland, especially during storms and high tides. (The rate of noticeable change is highly variable along the state's seismically active coastline. For instance, where local land surfaces are tectonically warped upward north of the Mendocino coast, sea-level rise has less impact.) All studies show that sea-level rise resulting from warmer seas will not only continue but accelerate over time. Furthermore, acidification is increasing as the oceans absorb some of the additional CO_2, threatening fisheries and productive rocky reefs and kelp forests. One of the many reasons scientists have labeled this new age the Anthropocene is that human-driven climate change is impacting every environment, every landscape, and every person. The natural cycles that shape the climate are still working in the background, but they are now being overshadowed by relatively shorter term human interference, which is forcing a very different kind of climate change than that caused by those cycles that have operated for millions of years.

More locally, human changes to the landscape, such as paving roads and channelizing rivers, exacerbate the impact of climate change. In cities around the world, from LA to London to Bengaluru, the urban heat island effect creates local pockets of heat that dissipate very slowly. However, because our global oceans continue warming and our remote arctic regions are warming fastest, we cannot explain increasing temperatures here or around the globe by blaming urban heat islands. For instance, separate studies during the last several years have shown that downtown San Francisco's average temperatures have been increasing at rates similar to those in the rest of the state and the world, even though the City's population has changed little since 1950. Likewise, the Bay Area's national parklands have experienced similar warming, though they have not been developed. While undeveloped parts of the state (such as north-central California) and remote locations (such as Death Valley and Needles) and every California region are warming, evidence shows it happening faster within our urban heat islands. So far, the South Coast wins in this temperature race with no real winners, while the northwest coast has been warming slowest.

The warming climate is already affecting people's day-to-day lives. By looking at data from the state's energy companies and NOAA, researchers can analyze how Californians use energy to cool and heat their homes. Measuring cooling and heating degree days is one clear indicator of the impact of climate change. It is no surprise that during recent decades, Californians have on average been using increasing amounts of energy and spending more money to cool their spaces on days above 65° and decreasing amounts of energy to warm their spaces on days cooler than 65°. This means more pull on the electrical grid, more money spent in each household, and more days each summer when, if a person doesn't have access to cooling technology, they risk serious health problems from overheating.

One of the most sweeping recently recognized impacts of climate change is associated with Earth's changing general circulation patterns. As the Arctic has warmed faster compared to most of the planet, researchers have noticed that larger-amplitude (more extreme loops in ridges and troughs) Rossby waves and jet streams in the upper atmosphere have stalled over our latitudes for longer periods. (These are the meandering high-level winds that impact the weather; see our pressure and winds discussion in the introduction.) You will remember how upper-level troughs of low pressure support and guide winter storms, while upper-level ridges of high pressure block them, producing fair, dry weather. When the atmospheric waves of troughs and ridges become more extreme, the intensity of weather extremes (hot vs. cold, wet vs. dry) increases at the surface and then prolongs these anomalies when the waves remain stationary. Many computer models have been predicting such wind circulation changes as nature redistributes the additional trapped heat within the atmosphere. A warmer atmosphere is also capable of absorbing more water vapor that could fuel wetter and more powerful storms when they form. Could this pattern of stalled waves in the atmosphere help explain California's extended record droughts that have been suddenly broken by flooding atmospheric rivers (ARs), in what seem to be erratic but repetitive cycles during the last few decades? Consider the floods of 2017 that broke a historic and prolonged drought and the floods of 2023 that broke another historic drought.

A clear trend continues to show a larger percentage of winter runoff flowing out of the state's mountain ranges from warmer storms that still bring snow, but at higher elevations than before. (The winter of 2023 was a stand-out exception to this trend.) On average, less annual precipitation is falling as snow, replaced by short-term downpours falling on lower mountain slopes. Such cloudbursts result in quick floods, as opposed to snowfall that stores water for longer periods and gradually releases this precious resource during the spring and summer months. In the Sacramento River watershed, average runoff from proportionally more rainfall and less snowfall has been flowing out with higher peak discharges and shorter lag times. These changes increasingly threaten water infrastructures and supplies during winter flood events and summer droughts. Further, with less snow cover, more of the sun's radiation is absorbed by darker surfaces, contributing to what scientists refer to as positive feedback. The result is accelerated warming, particularly on the microclimate scale.

Another aspect of climate change offers more evidence of complicated interactions between the atmosphere and oceans and the enormous amounts of energy exchanged within the Pacific Basin. Shorter-term (year-to-year) global temperature undulations may not match Golden State trends, possibly because El Niño periods tend to be stormy here, particularly in Southern California. Could the state's cloudier and wetter El Niño conditions suppress temperatures locally for a year or so until more stable high pressure and sunshine return to bring soaring temperatures?

Although temperatures have been creeping higher throughout most of the state, it is easier to notice local impacts within ecosystems. Numerous plant and animal species (many featured in previous chapters) have already been migrating up to 500 feet (152 m) higher in elevation throughout the Sierra Nevada and other mountain ranges. Recent research also suggests that marine layers have become less reliable sources of relief from summer drought. Warmer ocean temperatures have helped fuel warmer air temperatures, resulting in air that gains a greater capacity to hold moisture in the vapor state. Add the complex changes in air masses and weather patterns previously discussed. Should coastal low clouds and fog form less frequently and burn off faster, ecosystems will receive less moisture and, with increased sun exposure, more quickly lose what water they have. Meanwhile, inland areas distant from marine layers are experiencing record heat during summer, further exacerbating drought conditions. Climate models indicate that these trends will continue, but will be punctuated by seasons or entire years that buck the trends.

California plant communities and entire ecosystems are being transformed on a massive scale by these climate extremes. In historically cooler and wetter Northern California and the Sierra Nevada, pine forests are being replaced by more drought-tolerant oak woodlands. In Southern California, areas occupied by pines are also decreasing; but so are those occupied by oaks, as they are being replaced by grasslands and other more xeric plant communities. Decreasing soil moisture (particularly during summer drought) is impacting terrestrial ecosystems across the state.

Researchers are using tree ring reconstructions and soil moisture estimates to compare how today's climate change and its impacts on ecosystems might compare to past disruptions. Thousands of sites have been studied in the western US so that reliable soil moisture estimates can be made back to the year 800. (Researchers are simultaneously measuring charcoal deposits that date back 1,400 years before the present.) Evidence suggests

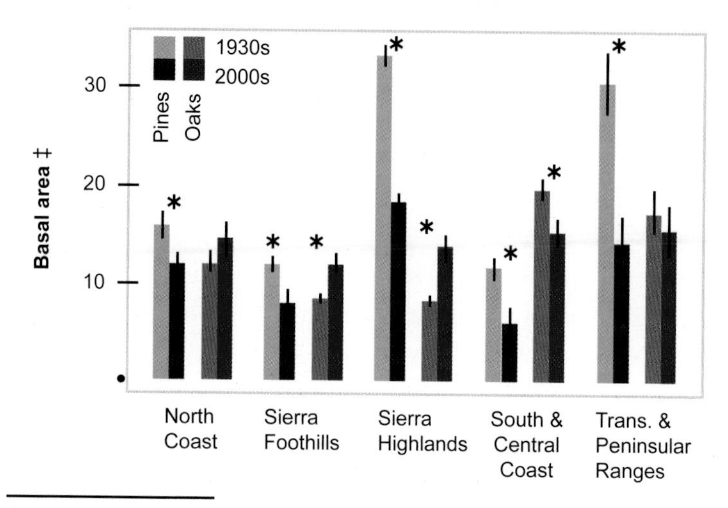

‡ Basal area refers to the area occupied by tree trunks

* Statistically significant differences

FIGURE 5-13. From north to south, drought-tolerant plant communities are replacing those requiring more water. *Source:* "Indicators of Climate Change in California." Sacramento: Office of Environmental Health Hazard Assessment, May 2018.

that the megadrought that started in California and the western US in 2000 was the worst in more than 1,200 years. Scientists estimate that 25–50 percent of this historic megadrought was human induced—caused by changes in atmospheric chemistry due to increased combustion of fossil fuels and other human activities, including deforestation and agricultural production—showing that an otherwise natural drought period was greatly exacerbated by human-driven climate change. There are some familiar players in this drought that crippled California: those resilient high-pressure systems blocked storms that would have otherwise watered our state. When winter storms finally swept through, they often brought brief, record-breaking downpours at higher elevations rather than snow, leading to faster runoff, flooding, and less water storage. These events were followed by record high temperatures; the dry, hot air sucked desperately needed moisture out of soil, plants, and even reservoirs. (The snowy winter and spring of 2023 are perfect examples of exceptional seasons that bucked those trends.)

HOW ARE CALIFORNIA'S LANDSCAPES AND PEOPLE REACTING AND ADAPTING TO CLIMATE CHANGE?

The latest megadrought, which dragged on through the first two decades of this century, devastated California's water systems and ecosystems. Central Valley farmers further depleted groundwater supplies, exhausting aquifers and causing the ground surface to sink, a growing worldwide problem known as subsidence. In our cities' urban forests, thousands of trees succumbed to drought, endangering shade canopies, which are more necessary than ever in a hotter climate. Skyrocketing water bills reflect rates raised higher for profligate water users. Consequently, attractive native plants and more xeric landscaping are replacing thirsty lawns, while parks and other public spaces use recycled water in their landscaping. Back in the 1970s, during the last big California drought, emptied pools attracted extreme skateboarders as skaters started doing tricks and flips up and down the sides of the gaping cement holes. What cultural transformations will bigger, longer droughts and severe floods bring? The most recent drought crisis has even plunged Californians into hot debates over proposed

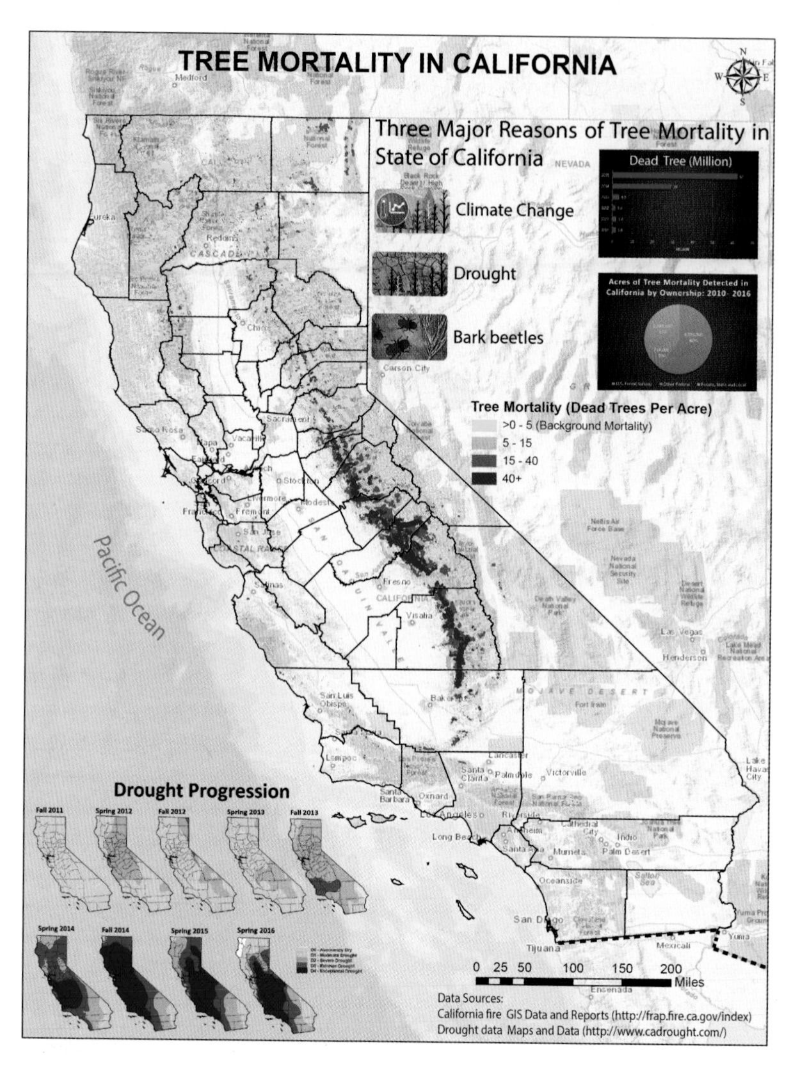

FIGURE 5-14. Tree mortality peaked (we hope) from 2010 to 2016, a period that included the great drought in the middle of our mega-drought. However, drought, bark beetles, and tree mortality rampaged through California forests (targeting mainly fir trees) again through 2022. *Source:* Map by Reza Mandalzadeh, as developed in Dr. Jing Liu's geographic information systems class.

desalination's exorbitant costs and its impacts on our marine ecosystems. One Californian's reasonable drought buster is another's douser and water witch. Although California has suffered from past years of drought punctuated by floods, the long-term data that researchers have been collecting, from tree rings to lake sediments to ice cores and ancient pollen samples, points in the same direction. And whether you think the solutions might include water conservation or cloud seeding, the primary causes are clear: human activities are changing our weather patterns and climates, thus disrupting the ability of natural systems and cycles to maintain equilibrium.

The most noticeable and devastating impacts of drought on California landscapes have come from wildfires. Although they were always major players in the state's ecology, wildfires more recently turned problematic and then calamitous. Forest mortality spread as tens of millions of trees were lost in individual years to drought and hungry bark beetles. The National Forest Service counted more than 200 million California trees that succumbed to drought and bark beetles from 2010 to 2022. That includes more than 62 million in 2016, the peak year. Another more than 36 million trees (particularly fir) died from renewed drought and bark beetle infestations in 2022. It is estimated that acreage burned in nonforest ecosystems increased by more than 190 percent, while forest acreage burned increased by more than 1,200 percent from the 1980s through the first two decades of the 2000s. This is likely the result of poor forest management and fire suppression that allowed fuels to accumulate since the late 1800s (a failed policy that ended by the late 1900s); the introduction of flammable, nonnative, invasive plants; and accelerating climate change. Add more people moving into and interacting with the wildland–urban interface, creating a complicated synergy of factors that have conspired to create deadly wildfire crises and chaos throughout California and other western states. As with the megadrought, a combination of paleoclimatology research and direct observations has helped identify these trends and their causes.

California is prone not just to megadroughts but to megastorms. This has been true since before we recognized climate change. A series of powerful winter storms swept through California in the winter of 1861–62 to produce our greatest historical natural disaster. ARs meandered and snaked across the West Coast from Oregon to Mexico, delivering alternating periods of cold, heavy mountain snows followed by warm, prolonged

FIGURE 5-15. Cycles of tree die-offs scarred many Sierra Nevada and other California landscapes as the megadrought dragged on through the first two decades of this century and summer soils dehydrated. Note the power lines crossing through the forest of dried fuels and the cabin hidden in a sea of dead and dying pines.

downpours. It is estimated that one-quarter to one-third of all property across the state was damaged or destroyed in floods that killed up to four thousand people (in a state with less than four hundred thousand at the time) and drowned several hundreds of thousands of farm animals. Rivers from Oregon to San Diego recorded their highest flood stages, some of which hold to this day. The flooding pooled until water covered an area 200 miles long and 20 miles wide in the Central Valley, while many other inland valleys were also submerged. Coastal floodplains lived up to their names. Cities went under for weeks as bridges and other infrastructure and entire towns were swept away. Recovery efforts lasted for many years, especially for the many Californians who had lost everything. The likelihood of aberrant and extreme weather events like this grows as climates change—bigger and wilder storms that visit with more frequency, causing flooding, such as when once-dormant Tulare Lake partially and temporarily reappeared during the 2023 record storms.

Existing climate models point toward more wild-weather anomalies and increasingly costly calamities on the long-term horizon. Just as unprecedented extreme weather events continue taking thousands of lives, doing billions of dollars of damage, and creating millions of refugees in each hemisphere every year, Californians are witnessing the impacts of climate change on their weather patterns during each season. Recent examples include when 2021 deteriorated into one of the worst water-shortage and fire years on record. By August, the majority of the state was suffering under exceptional to extreme drought. The Golden State's legendary water wars amped up again. And these water headaches extended far beyond our borders and across the West as the first water shortage ever declared in the Colorado River Basin was forcing mandatory water cuts in nearby states. Dehydrated ecosystems parched by brutal heat waves were burning in conflagrations that included the largest single wildfire and second-largest overall fire in California history.

Suddenly, two record-breaking AR events swept through in October and December 2021. As some of the state's rivers quickly swelled to flooding and reservoirs began to fill again, record mountain snows piled higher. And then nature's faucet suddenly shut off, right in the middle of California's traditional rainy season. The first three months of 2022 turned out to be the driest January–March in the state's recorded history. By summer, the drought was back on full blast, and so were the state's water woes. Those giant high-altitude ridges in the Rossby waves were stuck again. Then, the end of 2022 ushered in a series of record-setting ARs that continued into January 2023. More record snows and floods destroyed entire towns. After a brief lull, a historic cold storm spun down the coast from Canada in late February 2023, leaving snow on beaches from Crescent City down to Oxnard and spreading hailstorms widely. That storm grew into a monster off Southern California, generating flooding rains and mountain blizzards, closing major highways, and breaking more records. And this storm was followed by another drenching round of ARs in March 2023. Once again, Golden State weather was the top national news story. The events I've just outlined represent only a two-year sample of roller-coaster weather whiplash that ends here only because we have to go to publication. But they are sounding too familiar in this state where the unprecedented has become the normal; you are likely to be adding new weather pattern aberrations and disasters to the list as we continue past 2023.

The relentless deadly and destructive ARs that swamped our state during the 2023 rainy season highlighted our uneasiness with recurring weather-event uncertainties. Tons of record snowfall buried Sierra Nevada high country. Stormy seas destroyed beach boardwalks and damaged harbors. Communities, farms, and major highways went underwater. As the rivers and pools of water swelled, officials wondered if this could have marked the end of the state's megadrought. By the end of the rainy season, some scientists speculated that it was just another miniature dress rehearsal for the future ARkStorm (a megastorm scenario summarized in the next paragraph). Climate scientists repeated the same message: anomalous and destructive weather patterns have always challenged and threatened life in the Golden State, but they are becoming more frequent and powerful due to climate change. It is difficult, if not impossible, to precisely anticipate the scope and power of such future weather disasters.

A little more than ten years ago, scientists at the US Geological Survey and other agencies joined together to address threats posed by such an ARkStorm scenario if it were to hit us today. Sediment analysis and other evidence now warns that such devastating storms and deluges may have recurrence intervals averaging around 150–200 years. Climate change threatens to reduce that interval. A strong El Niño year coinciding with already warmer ocean water could provide catastrophic energy and moisture to potent storms. Such an event today could force evacuations of up to 1.5 million people and cause more than $700 billion in damage, even eclipsing the destruction from the state's anticipated major earthquake. One big difference from earthquake dangers is that scientists and public officials are getting better at forecasting and planning days ahead of time for the arrival of these ARs and their floods. Warning systems are becoming more reliable, thanks to an improved understanding of short- and long-term seasonal variations and rhythms in the atmosphere and oceans. But the increasingly ominous wild card is climate change.

If I had written this book before the 1980s, it would have ended several pages ago. That was the before times. We knew that human activities such as burning fossil fuels were polluting the air, but these were usually considered shorter-term problems affecting relatively local populations, and most had shorter-term local solutions. Cities were cleaning up car exhaust here and dirty power plants there so that people living around those cities could breathe healthy air again and avoid getting sick. When such efforts failed, dilution was the solution to pollution, as rain or wind would eventually clear it away. But as researchers made more measurements, their results revealed a global problem resembling a cancer that was gradually changing the chemistry of our entire atmosphere. Now we know that the changes are accelerating every year and that the effects will linger for many decades even if we were to stop polluting. I wish this weren't true, since it would make life a lot simpler and more happy-go-lucky without another problem dogging us as we plod along. Our dysfunctional human family would have a lot less to fight about without this additional cloud hanging over us. But instead of seeing this ongoing atmospheric experiment as a burden, why not consider it an existential opportunity created by an existential conundrum?

Uncertainty makes our challenge more difficult. Asking me (or anyone) to predict the precise temperature at any location in future decades or exactly how climate change will make you feel in a few years or decades is like a person asking what health problems and symptoms they will have if they keep smoking. We know the answer is likely to be disturbing, but who can know the exact details? The best answer is to start with the extreme weather events you see or hear about on the news every year, the ones that inspire disaster movies—and then multiply. Regardless of the specifics, all the computer models and projections predict a future with more numerous severe and anomalous events, and they will require a lot more preparation.

In chapters 1–4, we explored the weather patterns common to each season across California. We saw how they have shaped the landscapes, ecosystems, and people of the Golden State. Those chapters form a solid foundation for this chapter, because we must understand the science behind the current forces that drive our weather systems before we can understand how they

FIGURE 5-16. Looking from LA's tallest building toward the Hollywood Hills, we notice the sea of development that forms a robust urban heat island. Haze (left) creeps in with the sea breeze to accentuate an inversion capping the shallow marine layer. More than 50 years ago, before the establishment of the Air Quality Management District and air-quality control programs, the smog was twice as bad.

FIGURE 5-17. A rare, but recently more common, desert wildfire torched this Joshua tree woodland near Cima in the Mojave Desert. Parched nonnative grasses and other introduced species have conspired with record drought and heat to threaten these treasured symbols of the state's high-desert landscapes.

FIGURE 5-18. This is how the Joshua tree woodland in the previous image appeared before the fire. Invasive ephemeral grass seeds wait to spring up after the next rain.

Dana Glacier

Photos: U.S. Geological Service, photo station ric046 (left); H. Basagic (right)

Conness Glacier

Photos: National Park Service, photo station Conness 5555 (left); H. Basagic (right)

FIGURE 5-19. Here's what just over 55 years of climate change did to retreating glaciers in the highest elevations of the Sierra Nevada. *Source:* Research compiled from the USGS and National Park Service as shown in California Office of Environmental Health Hazard Assessment, "Glacier Change," February 11, 2019, https://oehha.ca.gov/epic/impacts-physical-systems/glacier-change.

are changing. How could we study climate change unless we have a starting point? What kind and what types of change are we talking about? Armed with our knowledge, we can see how the expectations and models built on reliable cycles of the past, the familiar weather patterns of our youth or that of the previous generation, are now erratic and shifting. And so our landscapes and ecosystems and people must respond to those changes. When such extreme weather patterns get locked in, they begin to resemble those familiar friends or relatives who overextend their stays and wear out their welcomes. And so it is with how climate change is driving many of our recent extreme, stubborn weather patterns. We must learn how to adjust and to cope with our guests.

HOW CAN WE RESPOND TO THE CHALLENGE?

The atmosphere and its interdependent heat, wind, moisture, and other weather elements operate on an almost incomprehensibly large scale. Yet we tiny humans have become major influencers in the complex systems and cycles that rule our atmosphere and our planet. We can make a collective decision to reconnect to these natural cycles and clean up our acts in ways that will benefit all. When we followed the science and used our technologies to identify pollution as the cause of the rapid loss of ozone in Earth's stratosphere, we discovered and implemented helpful solutions and made lifesaving progress in just a few decades. These were obvious life-and-death reactions and policy decisions. Similarly successful solutions must be sought that could slow the accelerating climate change in the troposphere (the atmospheric layer closest to the Earth's surface). Each year, hundreds of lives and billions of dollars in damage—caused by catastrophic floods, droughts, fires, deadly heat waves, and other severe weather events—are at stake in California alone. And this issue is not only about monetary loss. Californians are frequently inflicted with the anxiety and feelings of helplessness that accompany the increasingly anomalous weather patterns we must experience. We mourn the loss of our beloved unique natural treasures that have made California world famous. A few decades ago, we could count on the heavy winter rains and summer fog to nurture the tallest trees in the world and other coastal forests growing beyond our redwoods. But the rain and fog are not the reliable friends they once were, overwhelming us one year and nearly disappearing the next. We could once count on seasonal snows to nurture the largest trees in the world. But unreliable Sierra Nevada winters and occasional record-heat summers are often the rule today, leaving our giant sequoias and other trees and entire mountain forests in jeopardy. We could look forward to the balmy days of autumn, but now we dread being smoked out by the autumn wildfires that scorch our plant communities. There is a growing sense of dread and disappointment that is impossible to measure, as if we were losing beloved family members and friends to some type of plague. It takes the fun out of severe weather events, making us worry that we might lose our water resources one year and then get burned or flooded out the next. Yet, as nature calls out with more warnings, we look the other way and keep living as though nothing has changed.

When I was a little California kid, we lived our lives as if there were inexhaustible natural resources, unless their financial costs were too high. There were only three billion people in the world. We had all the well water we wanted. We turned off lights when we weren't using them only to save money. When I was old enough to drive and buy my first car, I considered the models with the best gas mileage (mileage was terrible compared to today's standards) so that I could get to work and school without falling into debt. There was no talk or consideration of global climate change. California's population (twenty million) was only about half of today's.

But things have changed. Those days are long gone. It's the end of that world and that California as we knew it. There are now more than eight billion of us on the planet. We have developed technologies that no person could have imagined, countless innovations that can destroy or save us. And most Americans require up to 25 acres of productive Earth surface per person each year to fuel their hyped-up consumer dopamine surges and to support their carbon footprints. It would require up to five Earths of natural resources if everyone on the planet lived like us Americans. Today, we realize how everything we do has an impact on others and on our future. As we live our everyday lives, we feel stuck in the system that's been built, so we wish for the arrival of a hero or some other miracle that will fix everything and save us. But your hero and savior, the one with the best answers and solutions to these challenges and problems, is looking back at you in the mirror every day. Whether you imagine us in a big bus headed for a cliff or on a big boat drifting toward a waterfall, the paradigm shift required to turn us around and toward a more promising future is in your reflection and in our collective refusal to stick with the status quo.

So you won't find all the solutions here. There's no way of winning when writing about climate change causes or solutions. This person just wants the facts. That person wants my opinion. Researchers want the latest hard data. Everyone seems to look for or possess the simple magic answer. But there isn't one. In this business, a little bit of hedging and uncertainty is not a weakness; it's a sign of honesty and integrity. And the solutions are just as complicated as the science. So I won't try to convince you or make you feel better or cover all the points you want made or make you believe there's nothing to worry about. One reader will think I'm overreacting; another will think I'm underreacting. Nevertheless, here are some suggestions.

Some of the solutions are very simple. Go back to the basics. Discover how a more practical, minimalist approach will help you get the maximum out of life. Think about what's most important in your life. It always has to start with fresh air, food, water, health, and shelter. They are all ecosystem services, and without them, you can't work or play or enjoy family and friends, follow your spiritual path, or even get up in the morning. If we expect to continue forward with eight billion–plus other people, we need that paradigm shift. So liberate yourself from all the background noise that keeps you on a daily treadmill to nowhere. Stop trying to swim upstream and start flowing with the nature that nurtures you. Reassess what makes you happy and how you can improve the quality of your life and the lives of those around you. Reexamine what it means to be successful in life and look for more meaning in your relationships with your surrounding environments and the people in them. These new paths will lead to solutions to climate change and other growing problems that threaten our existence. Then, prepare yourself for the extreme weather events that are in our future, while doing what you can to keep them from getting worse. The stakes are too high to ignore the golden opportunity that presents itself. Take it. Everyone will be a lot better off. Folks reading this many years from now will thank us.

There are also plenty of broader policy changes that will lead us toward the light. Instead of relying on distant energy sources shipped hundreds of miles for your consumption, imagine a sensible power grid with local sources. Build cities that encourage safe and healthy walking and bike riding. Start a walking school bus in your community to get your kids to school safely on foot. Insist on more efficient vehicles and appliances that will save all of us a lot of money in the long run. Encourage your neighbors to get off *their* treadmills to nowhere. Your family, friends, and neighbors will jump on this wave that will build into a tsunami when they recognize the long-term benefits to them. Invite them on this train heading toward a future so bright, you're gonna need shades.

We hear and read about the big public policy responses debated on the news every day. There are countless California examples of actions and policies that will make positive change, such as along our waterways. Taking down obsolete dams that have lost their relevance (such as in Malibu Canyon and along the Klamath River) is a perfect example. This will allow salmon and steelhead (see chapter 4) to return to these streams and bring nutrients back, reintroduce natural cycles that will restore their ecosystems, and rejuvenate fishing industries that have been crippled by climate change and other human impacts. Building and maintaining debris basins within existing flood control channels will protect thousands of downstream California neighborhoods from deadly and costly floods and debris flows as severe events become more common. Extending water spreading grounds in river channels encourages runoff to soak in and recharge groundwater supplies before reaching the ocean. Restoring riparian habitats along our urban river channels will nurture native species and cool the urban heat island. We're doing all of these things right now, and we need to do more. Understanding and working with (instead of against) natural systems and cycles are our best insurance to prevent certain future suffering and catastrophes that will result if we bury our heads in the very sand that was deposited by California's rivers. Smart short-term investments now will produce long-term payoffs.

Imagine if we were creatures from another solar system visiting Earth for the first time. Upon our arrival, we would notice the jaw-dropping beauty of this little blue planet. But we might also find humans' absurdly destructive habits totally insane: single drivers go into debt buying the fuel required to lug tons of gargantuan SUVs around their cities; toilet paper is made from precious forests instead of from recycled products; people walk away from unfinished plates of food instead of sharing what they can't eat or taking it home for tomorrow's leftovers, wasting resources and energy and polluting the air just to have food delivered that is only thrown out; bales of alfalfa are grown with California's groundwater and shipped out to feed cattle across the world; people drink water in plastic bottles imported from across the Pacific Ocean, generating emissions both in making the bottles and in shipping them. It's pretty easy to extend this list in our fast-fashion culture track that seems to have fifty-two seasons in a year.

So there are also countless examples of low-hanging fruit waiting and calling out to us, simple tools in a handy waste-not, want-not toolbox that will slow climate change and the depletion of resources. My students and colleagues at Santa Monica College taught me to never give up. Years ago, we successfully lobbied and worked with our leaders at the college and in the city government to make our community more pedestrian-, bicycle-, and public transportation–friendly, until the majority of students and employees no longer commuted in their private vehicles. These efforts keep tons of pollution out of the air each week, ease a lot of traffic congestion, and make our neighbors happy. To set a fun and personal example of efficiency, I transformed my yard into a wild native landscape more than two decades ago. My California ornamentals provide native habitat and save water and the energy required to pump it.

All of the solutions suggested in the preceding paragraphs involve changes that make us more efficient. They represent the antithesis of sacrifice. They are actions that can simultaneously enhance the quality of our individual lives, clean up our communities, and heal the planet. The only thing controversial about them is that someone might consider them controversial. We are reaching the end of this book, but you have been launched into the next volume, one that will be written by you. The stories and highlights will range from individual small-scale actions to collective global-scale policies as you look toward an uncertain future. We can at least start with those easiest, most prudent, practical, beneficial adjustments while we continue healthy, open debates about our public policy directions and decisions.

Our challenge is to reject the short-term illusions that trick us into trying to conquer and then ignore nature. Earth's natural rhythms, cycles, and systems will always rule our lives in the long run. Our recognition and understanding of these dominating natural processes and powerful forces will determine our future success on this planet. Will we continue trying to row against the current with our blinders on and suffer the consequences, or will we learn to go with nature's flow? Our mental and physical health and our very survival will be determined by our answers to these questions.

FIGURE 5-20. You will find some of the most photographed badlands in the world at Zabriskie Point in Death Valley. A harsh climate shapes a vulnerable terrain. We are fortunate that such barren landscapes do not dominate the Golden State or the future.

EPILOGUE

Our lives are sometimes more similar to radically changing weather than we might admit. One day you are floating on air; next day you're being crushed by the weight of the world on your shoulders. The cycles of changing weather patterns echo the cycles in our lives, constantly reminding us how each day offers new opportunities to view our world and our place in it with fresh perspectives. Monotonous, dull gray periods with obstructed views clear into surprisingly brilliant, crystal-clear skies with unlimited visibility. Just when you have been lulled into complacency by the calm, winds of change sweep through, bringing bright sunshine and refreshing clarity or unsettling turbulence. Illusions of unchanging safety and security are blown away by the realities, challenges, and dangers that help define unstable weather and climate and life on our planet.

Every time I walk outside, I am inspired by the Earth's sustenance and beauty that surrounds and connects us all. What did you get out of this book, and where does it lead you from here? I hope it refreshed and reenergized your senses, helped to improve the quality of your life experience, and bolstered your understanding of our world. Ultimately, we can either wither behind our four walls, trapped in our artificial bubbles, or pull back the curtain to reveal the magnificence and magic and science waiting right outside our windows. Nature's ongoing experiments and lessons about real life in the real world are beckoning all around us. The air surrounding and nurturing you and the sky dome above you are calling out in one spectacular free motivational drama after another. If our atmosphere produces the greatest shows on Earth, California may be the greatest and most diverse and thrilling theater as we explore and learn more in our constantly changing wild-weather rides. Keep your seat belt fastened.

ACKNOWLEDGMENTS

Thanks to the folks at the National Weather Service for greeting and educating my students over the decades. More recently, Todd Hall and other forecasters became invaluable educators by giving their precious time to students at their Oxnard office. The many themed workshops and seminars they organized and led throughout the years for educators who teach atmospheric science and weather and climate courses have been very informative. Same goes to all the forecasters in NWS offices across the state for sharing their knowledge and expertise. We have all benefited from their efforts, especially me, and now you, as you explore this book.

I am so grateful to the thousands of students, professors, and colleagues from Santa Monica College and dozens of other colleges and universities who shared their research and perspectives during our field classes, labs, seminars, and class sessions. Together, we explored and researched nearly every landscape, weather pattern, climate, and learning environment imaginable. Special thanks to my colleagues in our Earth Science Department, biologists from Life Science, our sustainability heroes, and scholars from a range of disciplines who accompanied us and invited us to join them on so many experiential learning adventures. I would name all of you if it didn't require an entire additional chapter. More thanks go to the scientists and interpreters at the National Park Service (such as rangers Susan Teel and Antonio Solorio), state parks, visitor centers, and museums, and to local park guides. I have also gained a wealth of information from research, paper presentations, conferences, and seminars sponsored by numerous professional organizations in geography and atmospheric science, such as the American Meteorological Society, the California Geographical Society, and the Association of Pacific Coast Geographers. Big thanks to the many regional docent programs, championed by dedicated volunteers; these are the heroes who dedicate their precious time introducing thousands of Californians to the magic of natural history and field science.

It seems absurd to try to thank the professionals at Heyday in one short paragraph. Marthine Satris is a one-of-a-kind senior acquisitions editor who spent more than two years reshaping this into a much-improved book. Michele Jones is the exceptional copyeditor with impeccable attention to detail, leaving no stone (or cloud) unturned. Managing editor Emmerich Anklam kept his overarching eye on our progress. Art director Archie Ferguson, along with Alan Dino Hebel and Ian Koviak at *the*BookDesigners, is responsible for the attractive style. And my gratitude extends to all the staff at Heyday for their support of this book.

To all my family and friends (especially Jim, Julie, Sotath, and Alison) who understood my fascination with weather over the decades and accompanied me on memorable nature adventures, many thanks for your shared enthusiasm. I've been very fortunate to have you in my life.

CALIFORNIA'S DIVERSE CLIMATES IN MAPS AND GRAPHS

DISCOVERING TRENDS IN OUR DIVERSE CLIMATES

Just as a picture can be worth thousands of words, so can a climograph display a wealth of information about climate trends that could never be shared with words alone. Robert O'Keefe has carefully and masterfully researched and analyzed the most recent climate data to create the following up-to-date climographs. You will notice the same general seasonal trends that have been established by climate records found in countless past studies and publications. But you may also detect some subtle changes at some of the stations, particularly in their classifications. For instance, Los Angeles has long been recognized in the Köppen system as the classic mild Mediterranean climate. But

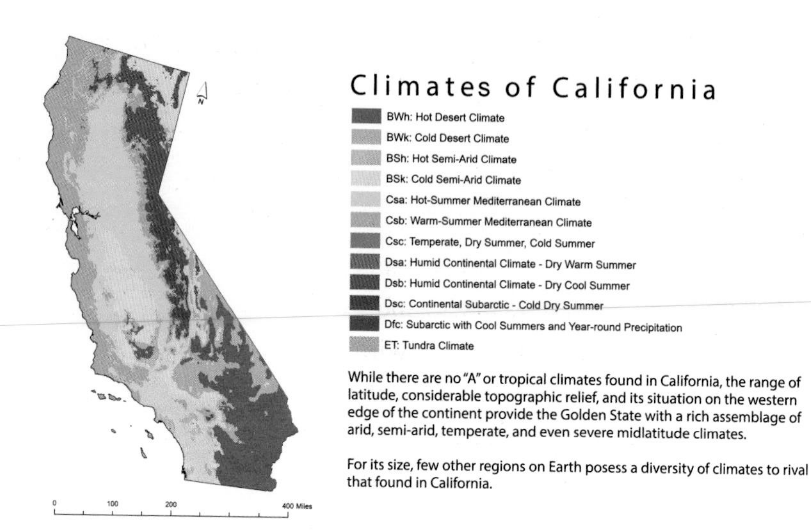

FIGURE A. Every attempt to classify California's diverse climates has its specific strengths and weaknesses. This is the best Köppen Climate map available at this scale. *Source:* Created by Robert O'Keefe.

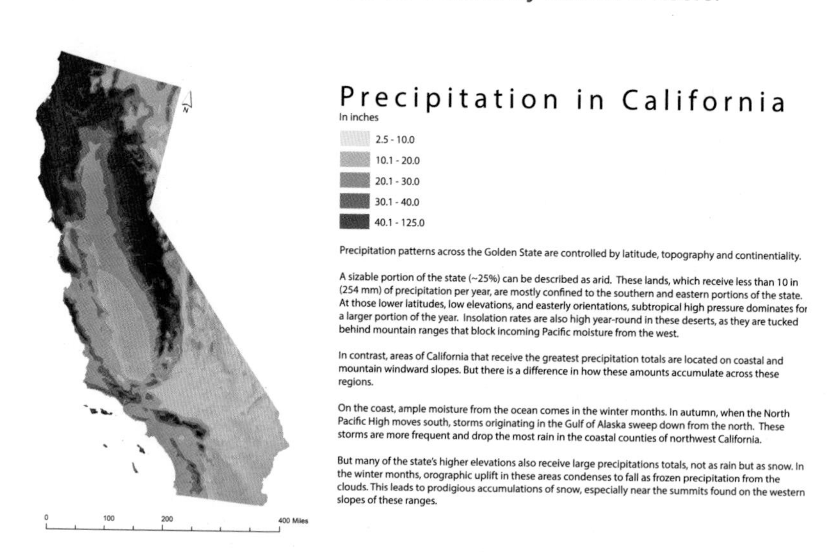

FIGURE B. California's average annual precipitation ranges from about 2 inches (5 cm) of rain in the driest southeastern low deserts to more than 100 inches (254 cm) on the wettest northwest slopes. This is the best average annual precipitation map available at this scale. *Source:* Created by Robert O'Keefe.

Rob notes how the first two decades of this century included historic warm and dry megadrought years that may have shifted LA into a more semiarid climate, at least temporarily. Several consecutive years of exceptional rainfall and cool temperatures could shift the classification back. But Rob is intent on showing the climate trends that we are experiencing today, rather than describing climates of the past, and we are dedicated to sharing the most recent facts and trends with you.

You may also notice some data that challenges your assumptions about climates at particular California locations. Note how temperature and precipitation totals are highly dependent on when the data was gathered and the microclimate at each particular location. For instance, after extensive research, we found reliable sources claiming that Lake Tahoe's average annual precipitation (including the amount of liquid water in the snowfall) totals were just over 16 inches on one side of the lake and up to 52 inches on mountain slopes above the other side! It turns out that not only were the years of record often different but there are a host of other factors in play that include elevation and slope exposure. And some of the stations may have only been near enough to Lake Tahoe to use the name. You can see why I've highlighted similar peculiarities within more general and familiar trends throughout this book, as California's incredibly diverse and changing climates call out for more attention.

1
2
3
4
5
6
7
8
9
10
11
12
13
14
15
16
17
18

1

Crescent City
Period of record: 1970-2012

Elevation: 14m/46'
Annual PPT: 70.97 in.
Mean temperature: 52.7°F
Climate Classification: Csb

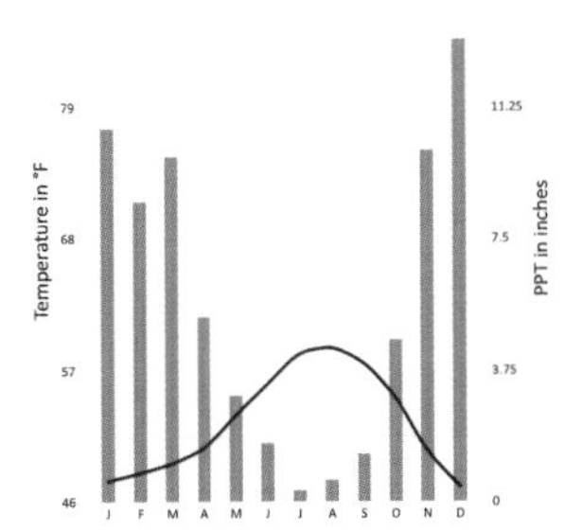

2

Eureka
Period of record: 1990-2020

Elevation: 13m/42.6'
Annual PPT: 39.98 in.
Mean temperature: 53.1°F
Climate classification: Csb

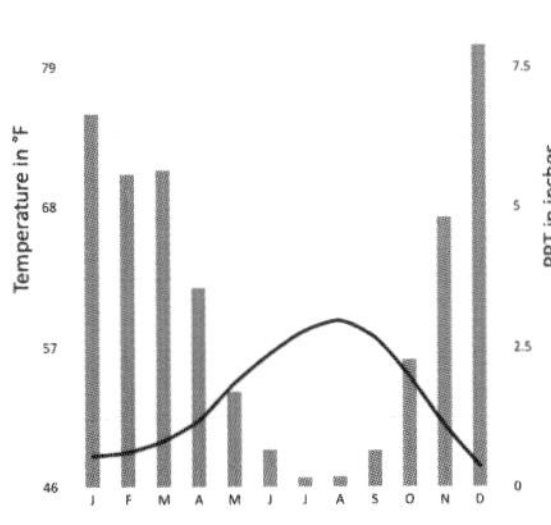
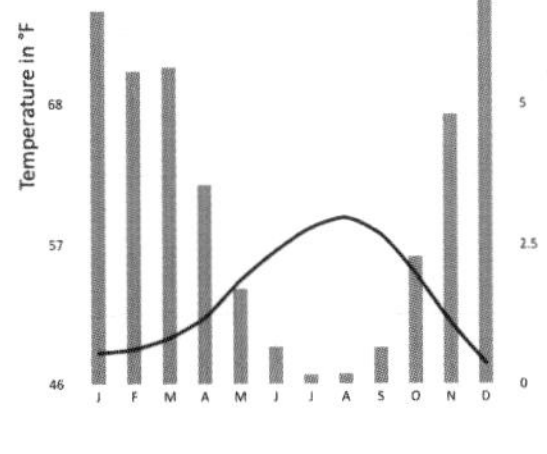

3

Redding
Period of record: 1990-2020

Elevation: 163m/537'
Annual PPT: 33.21 in.
Mean temperature: 63.0°F
Climate Classification: Csa

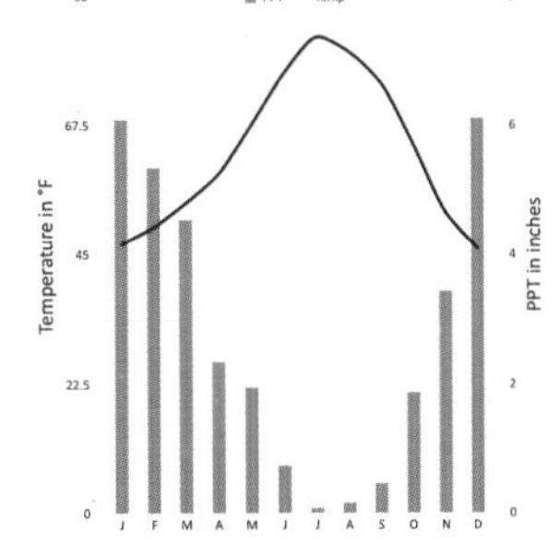

4

Alturas
Period of record: 1990-2020

Elevation: 1,333m/4,373'
Annual PPT: 10.74 in.
Mean temperature: 50.7°F
Climate Classification: BSk

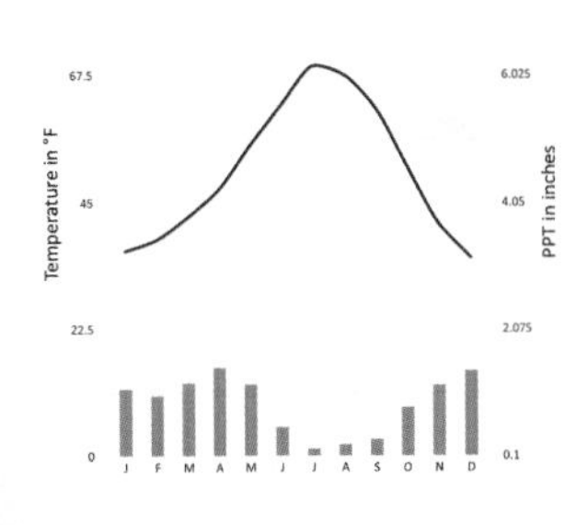

5

San Francisco
Period of record: 1990-2020

Elevation: 46m/151'
Annual PPT: 22.57 in.
Mean temperature: 58.2°F
Climate classification: Csb

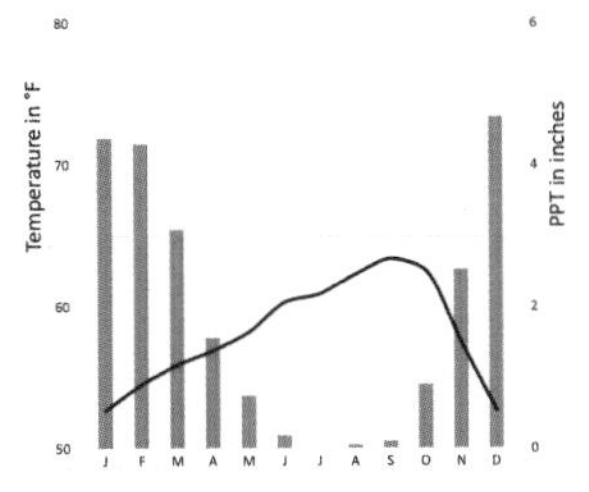

6

San Jose
Period of record: 1990-2020

Elevation: 25m/82'
Annual PPT: 13.51 in.
Mean temperature: 60.9°F
Climate Classification: Csb

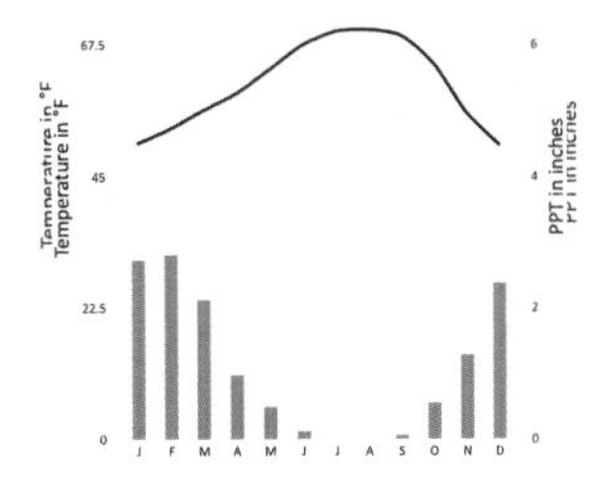

7

Santa Cruz
Period of record: 1990-2020

Elevation: 6m/19.6'
Annual PPT: 30.25 in.
Mean temperature: 58.2°F
Climate classification: Csb

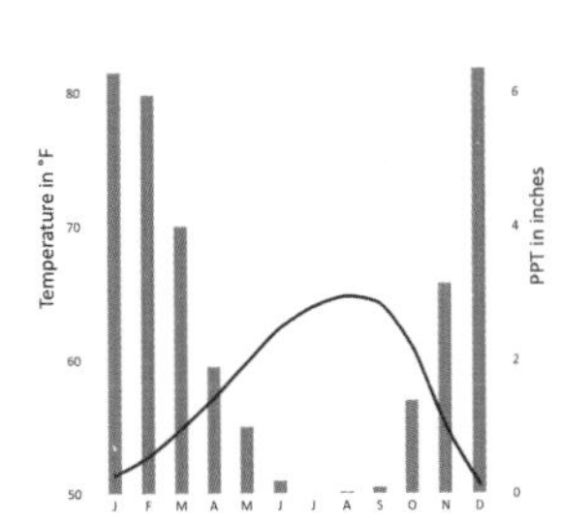

8

Sacramento
Period of record: 1990-2020

Elevation: 8m/26'
Annual PPT: 19.05 in.
Mean temperature: 63.8°F
Climate Classification: Csa

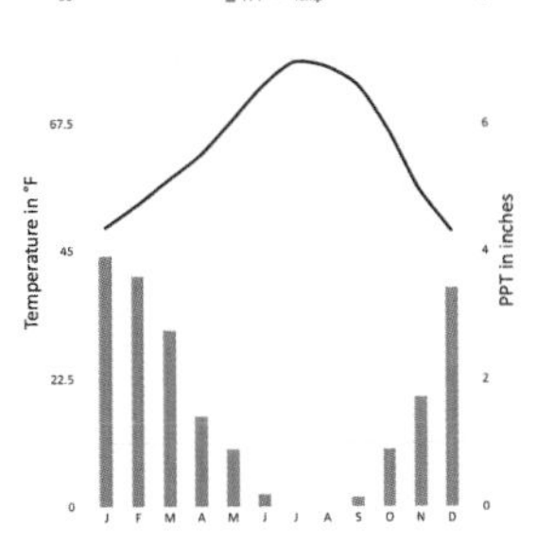

9

Truckee Ranger Station
Period of record: 1904-2016

Elevation: 1,773m/5,817'
Annual PPT: 30.15 in.
Mean temperature: 43.45°F
Climate classification: Dsc

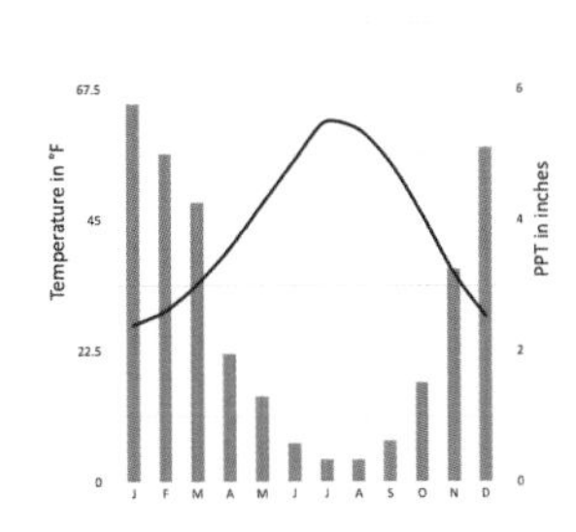

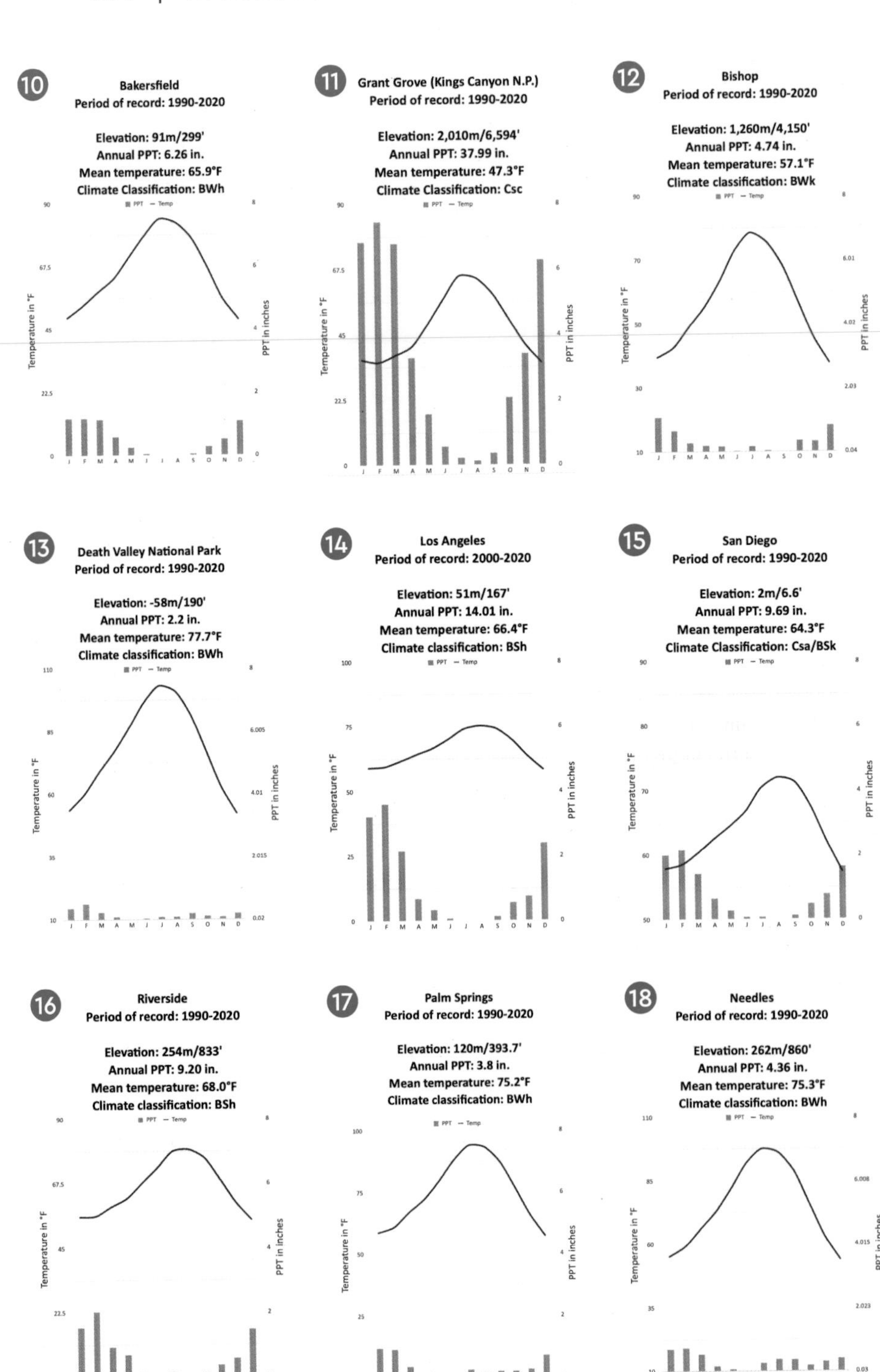

10
Bakersfield
Period of record: 1990-2020
Elevation: 91m/299'
Annual PPT: 6.26 in.
Mean temperature: 65.9°F
Climate Classification: BWh
PPT — Temp
Temperature in °F
PPT in inches
J F M A M J J A S O N D
11
Grant Grove (Kings Canyon N.P.)
Period of record: 1990-2020
Elevation: 2,010m/6,594'
Annual PPT: 37.99 in.
Mean temperature: 47.3°F
Climate Classification: Csc
PPT — Temp
Temperature in °F
PPT in inches
J F M A M J J A S O N D
12
Bishop
Period of record: 1990-2020
Elevation: 1,260m/4,150'
Annual PPT: 4.74 in.
Mean temperature: 57.1°F
Climate classification: BWk
PPT — Temp
Temperature in °F
PPT in inches
J F M A M J J A S O N D
13
Death Valley National Park
Period of record: 1990-2020
Elevation: -58m/190'
Annual PPT: 2.2 in.
Mean temperature: 77.7°F
Climate classification: BWh
PPT — Temp
Temperature in °F
PPT in inches
J F M A M J J A S O N D
14
Los Angeles
Period of record: 2000-2020
Elevation: 51m/167'
Annual PPT: 14.01 in.
Mean temperature: 66.4°F
Climate classification: BSh
PPT — Temp
Temperature in °F
PPT in inches
J F M A M J J A S O N D
15
San Diego
Period of record: 1990-2020
Elevation: 2m/6.6'
Annual PPT: 9.69 in.
Mean temperature: 64.3°F
Climate Classification: Csa/BSk
PPT — Temp
Temperature in °F
PPT in inches
J F M A M J J A S O N D
16
Riverside
Period of record: 1990-2020
Elevation: 254m/833'
Annual PPT: 9.20 in.
Mean temperature: 68.0°F
Climate classification: BSh
PPT — Temp
Temperature in °F
PPT in inches
J F M A M J J A S O N D
17
Palm Springs
Period of record: 1990-2020
Elevation: 120m/393.7'
Annual PPT: 3.8 in.
Mean temperature: 75.2°F
Climate classification: BWh
PPT — Temp
Temperature in °F
PPT in inches
J F M A M J J A S O N D
18
Needles
Period of record: 1990-2020
Elevation: 262m/860'
Annual PPT: 4.36 in.
Mean temperature: 75.3°F
Climate classification: BWh
PPT — Temp
Temperature in °F
PPT in inches
J F M A M J J A S O N D

SOURCES AND RECOMMENDATIONS FOR FURTHER READING

THE SOURCE OF YOUR FORECASTS

The National Weather Service (NWS) is part of the National Oceanic and Atmospheric Administration (NOAA), which is a branch of the Department of Commerce. Virtually all US weather forecasts, maps, charts, and satellite images (including the ones you see in this publication and project) originate from the NWS. As previously mentioned, all of these treasures are free and accessible to everyone, since we all paid for them.

Some public agencies and private firms add their own information and technologies to create more specific forecasts for clients (such as in agriculture, transportation, utilities, emergency management, tourism, and entertainment). Forecasts on TV, radio, and your favorite app omit or add specific information and graphics to the NWS forecasts to attract attention and increase ratings in highly competitive media environments. This currently includes the Weather Company (Weather Underground), the Weather Channel, and other various broadcasts and apps.

Private companies have experienced tumultuous upheavals over the years, including mergers, buyouts, cost-cutting, layoffs, and cancellations driven by ratings and short-term profits. Because it is impossible to predict whether these evolving or devolving businesses will be emphasizing information and education or entertainment when you read this, it is best to stick with the original and most stable and dependable source: the National Weather Service, www.weather.gov.

RECOMMENDED READING

BOOKS

Ahrens, C. Donald. *Meteorology Today*. Belmont, CA: Brooks/Cole, 2013.

Ahrens, C. Donald, and Robert Henson. *Meteorology Today: An Introduction to Weather, Climate, and the Environment*. 13th ed. Boston: Cengage Learning, 2022.

Austin, Mary. *The Land of Little Rain*. Boston: Houghton Mifflin, 1903.

Baily, Harry P. *Weather of Southern California*. Berkeley: University of California Press, 1966.

Bakker, Elna S. *An Island Called California*. 2nd ed. Berkeley: University of California Press, 1984.

Carle, David. *Introduction to Air in California*. Berkeley: University of California Press, 2006.

Carson, Rachel. *Silent Spring*. Boston: Houghton Mifflin, 1962.

Cunningham, Laura. *A State of Change: Forgotten Landscapes of California*. Berkeley, CA: Heyday, 2010.

Dale, Nancy. *Flowering Plants of the Santa Monica Mountains*. Sacramento: California Native Plant Society, 2000.

Dana, Julian. *The Sacramento: River of Gold*. New York: Farrar and Rinehart, 1939.

Felton, Earnest I. *California's Many Climates*. Palo Alto, CA: Pacific Books, 1965.

Gilliam, Harold. *Weather of the San Francisco Bay Region*. Berkeley: University of California Press, 2002.

Harden, Deborah R. *California Geology*. Upper Saddle River, NJ: Pearson Education, 2004.

Heizer, Robert F., and Albert B. Elsasser. *The Natural World of the California Indians*. Berkeley: University of California Press, 1980.

Hood, Charles. *A Californian's Guide to the Birds among Us*. Berkeley, CA: Heyday, 2017.

Kaufmann, Obi. *The California Field Atlas*. Berkeley, CA: Heyday, 2017.

Maclean, John N. *The Esperanza Fire*. Berkeley, CA: Counterpoint, 2013.

Marianchild, Kate. *Secrets of the Oak Woodlands*. Berkeley, CA: Heyday, 2014.

Merenlender, Adina, and Brendan Buhler. *Climate Stewardship: Taking Collective Action to Protect California*. Berkeley: University of California Press, 2021.

Mooney, Harold, and Erika Zavaleta. *Ecosystems of California*. Oakland: University of California Press, 2016.

Muir, John. *The Mountains of California*. American Museum of Natural History, 1981. Originally published 1894.

Pretor-Pinney, Gavin. *The Cloudspotter's Guide*. Somerton, United Kingdom: Cloud Appreciation Society, 2006.

Schoenherr, Alan A. *A Natural History of California*. Oakland: University of California Press, 2017.

Selby, William A. *Rediscovering the Golden State: California Geography*. 4th ed. Hoboken, NJ: Wiley, 2018.

JOURNALS

Bulletin of the American Meteorological Society: https://journals.ametsoc.org

California Geographer (journal of the California Geographical Society): http://calgeog.org/journal

Weatherwise: https://www.tandfonline.com/journals/vwws20

Yearbook of the Association of Pacific Coast Geographers: https://muse.jhu.edu/journal/280

NEWSPAPERS

Here are some of the largest California newspapers with histories of weather stories and forecasts. Smaller-town local newspapers have also recorded a wealth of weather events since the 1800s.

Fresno Bee: https://www.fresnobee.com/

Los Angeles Times: https://www.latimes.com/

Sacramento Bee: https://www.sacbee.com/

San Diego Union-Tribune: https://www.sandiegouniontribune.com/

San Francisco Chronicle: https://www.sfchronicle.com/

San Jose Mercury News: https://www.mercurynews.com/

ORGANIZATIONS, AGENCIES, AND SOME SELECT WEBSITES

Adaptation Clearinghouse: https://www.adaptationclearinghouse.org/

American Meteorological Society: https://www.ametsoc.org/index.cfm/ams/about-ams/

Association of Pacific Coast Geographers: http://www.apcgweb.org/

Cal-Adapt: https://cal-adapt.org/

California Climate Change Center: https://www.adaptationclearinghouse.org/organizations/california-climate-change-center-cccc.html

California Climate Data Archive: "Welcome to Calclim, the California Climate Data Archive." https://calclim.dri.edu/.

California Coastal Commission: https://coastal.ca.gov/

California Coastal Commission—Climate Change: https://www.coastal.ca.gov/climate/climatechange.html

California Department of Water Resources: https://water.ca.gov/

California Energy Commission: https://www.energy.ca.gov/

California Extreme Precipitation Symposium: https://cepsym.org/

California Geographical Society: http://calgeog.org/

California Governor's Office of Planning and Research (Climate Risk and Resilience): https://opr.ca.gov/

California Historical Society: https://californiahistoricalsociety.org/

California Native Plant Society: https://www.cnps.org/

California Natural Resources Agency: https://resources.ca.gov/

California State Coastal Conservancy: https://scc.ca.gov/

California Water Science Center: https://www.usgs.gov/centers/california-water-science-center

Center for Climate Change Impacts and Adaptation: https://climateadapt.ucsd.edu/

Center for Western Weather and Water Extremes: https://cw3e.ucsd.edu/

Climate records for our big cities: https://www.climatestations.com/

Cloudman: https://www.cloudman.com/

Desert Research Institute: www.dri.edu

Golden Gate Weather Services: https://ggweather.com/

Intergovernmental Panel on Climate Change: https://www.ipcc.ch/

JPL Atmospheric Infrared Sounder: https://airs.jpl.nasa.gov/

Libraries scattered across the state (from local small towns to the state library) house countless weather-related publications and data, including historical accounts of weather events. Same goes for libraries in many of our schools, colleges, and universities.

Monterey Bay Aquarium: https://www.montereybayaquarium.org/

NASA/National Lightning Detection Network: https://ghrc.nsstc.nasa.gov/home/lightning/index/data_nldn

National Science Foundation: https://www.nsf.gov/

National Science Foundation: Climate Change—Special Report: https://nsf.gov/news/special_reports/climate/

National Science Teaching Association: https://www.nsta.org/

National Science Teaching Association Climate Change Resources: https://www.nsta.org/topics/climate-change

National Weather Service Climate Prediction Center: https://www.cpc.ncep.noaa.gov/

National Weather Service ENSO discussion: https://www.cpc.ncep.noaa.gov/products/precip/CWlink/MJO/enso.shtml

National Weather Service History in California: https://www.weather.gov/media/ilx/History/california_wb.pdf

National Weather Service Los Angeles/Oxnard, senior meteorologist Todd Hall and the entire crew: https://www.weather.gov/lox/

National Weather Service offices and regions: https:www.weather.gov/srh/nwsoffices

National Weather Service San Diego, *Coast to Cactus Weather Examiner*: https://www.weather.gov/media/sgx/newsletter/current-newsletter.pdf

NOAA Climate.gov: https://www.climate.gov/search?query=california

NOAA National Centers for Environmental Prediction, Weather Prediction Center: https://www.wpc.ncep.noaa.gov/dailywxmap/index.html

NOAA Ocean Prediction Center—Pacific Analysis: https://ocean.weather.gov/Pac_tab.php

Pacific Decadal Oscillation (PDO): https://www.ncdc.noaa.gov/teleconnections/pdo/

PG&E, "The Pacific Energy Center's Guide to: California Climate Zones": https://www.pge.com/includes/docs/pdfs/about/edusafety/training/pec/toolbox/arch/climate/california_climate_zones_01-16.pdf

Prism Climate Group at Oregon State University: http://prism.oregonstate.edu

Scripps Institute of Oceanography, "FAQ: Climate Change in California": https://scripps.ucsd.edu/

Scripps Institute of Oceanography Climate Sciences: https://scripps.ucsd.edu/research/topics/climate-sciences

Sunset Climate Zones: https://www.sunsetwesterngardencollection.com/climate-zones

Swain, Daniel. *Weather West*. Unique perspectives and updates on California weather and climate from a climate scientist: https://weatherwest.com/

University of California: https://www.universityofcalifornia.edu/subject/term/climate-change

University of California Division of Agriculture and Natural Resources Climate Zones: https://cagardenweb.ucanr.edu/Your_Climate_Zone/

US Department of Agriculture, California USDA Plant Hardiness Zone Map: https://www.plantmaps.com/interactive-california-usda-plant-zone-hardiness-map.php

US Drought Monitor: https://droughtmonitor.unl.edu/C3urrentMap/StateDroughtMonitor.aspx?CA

US Geological Survey. Research such as the USGS Multi-Hazard West Coast Winter Storm Project, 2011; Keith Porter et al., "Overview of the ARkStorm Scenario," June 7, 2011, https://pubs.usgs.gov/of/2010/1312/; and Science Application for Risk Reduction, "ARkStorm Scenario," January 23, 2018, https://www.usgs.gov/programs/science-application-for-risk-reduction/science/arkstorm-scenario

Western Regional Climate Center: https://wrcc.dri.edu/

World Meteorological Organization: https://public.wmo.int/en

RELEVANT RESEARCH, JOURNALS, AND ARTICLES: A SELECT BIBLIOGRAPHY

Abatzoglou, John T., Kelly T. Redmond, and Laura M. Edwards. "Classification of Regional Climate Variability in the State of California." *Journal of Applied Meteorology and Climatology* 48, no. 8 (2009): 1527–41.

Blier, Warren. "The Sundowner Winds of Santa Barbara, California." *Weather and Forecasting* 13, no. 3 (1998): 702–16. https://doi.org/10.1175/1520-0434(1998)013<0702:TSWOSB>2.0.CO;2.

California Department of Water Resources. *California's Most Significant Droughts: Comparing Historical and Recent Contributions*. California Department of Water Resources, 2015. https://cawaterlibrary.net/wp-content/uploads/2017/05/CalSignficantDroughts_v10_int.pdf.

California Natural Resources Agency. *Safeguarding California Plan: 2018 Update—California's Climate Adaptation Strategy*. California Natural Resources Agency, 2018. https://files.resources.ca.gov/docs/climate/safeguarding/update2018/safeguarding-california-plan-2018-update.pdf.

California Office of Environmental Health Hazard Assessment. "Glacier Change." *CA.gov*, February 11, 2019. https://oehha.ca.gov/epic/impacts-physical-systems/glacier-change.

California Office of Environmental Health Hazard Assessment. "Indicators of Climate Change in California." CA.gov, 2018. https://oehha.ca.gov/climate-change/document/indicators-climate-change-california.

Chenoweth, Michael, and Christopher Landsea. "The San Diego Hurricane of 2 October 1858." *Bulletin of the American Meteorological Society* 85, no. 11 (2004): 1689–97. https://www.jstor.org/stable/26221214.

Choi, Jung, and Seok-Woo Son. "Seasonal-to-Decadal Prediction of El Niño–Southern Oscillation and Pacific Decadal Oscillation." *npj Climate and Atmospheric Science* 5, no. 29 (2002). https://www.nature.com/articles/s41612-022-00251-9.

Dettinger, Michael D. "Atmospheric Rivers as Drought Busters on the U.S. West Coast." *Journal of Hydrometeorology* 14 (2013): 1721–32.

Dettinger, Michael D. "Climate Change, Atmospheric Rivers, and Floods in California—A Multimodel Analysis of Storm Frequency and Magnitude Changes." *Journal of the American Water Resources Association* 47 (2011): 514–23.

Gonzalez, Patrick. *Climate Change in the National Parks of the San Francisco Bay Area, California, USA*. National Park Service, 2016. http://www.patrickgonzalez.net/images/Gonzalez_climate_change_SF_NPS.pdf.

Grubišić, Vanda, and John M. Lewis. "Sierra Wave Project Revisited." *Bulletin of the American Meteorological Society* 85, no. 8 (2004): 1127–42. https://journals.ametsoc.org/view/journals/bams/85/8/bams-85-8-1127.xml.

He, Minxue, Michael Anderson, Andrew Schwarz, Tapash Das, Elissa Lynn, Jamie Anderson, Armin Munévar, Jordi Vasquez, and Wyatt Arnold. "Potential Changes in Runoff of California's Major Water Supply Watersheds in the 21st Century." *Water* 11, no. 8 (2019): 1651. https://doi.org/10.3390/w11081651.

He, Minxue, and Mahesh Gautam. "Variability and Trends in Precipitation, Temperature and Drought Indices in the State of California." *Hydrology* 3, no. 2 (2016): 14. https://www.mdpi.com/2306-5338/3/2/14.

Kaplan, Curt, and Rich Thompson. "The Palmdale Wave: An Example of Mountain Wave Activity on the Lee Side of the San Gabriel Mountains." Weather Service Forecast Office Los Angeles/Oxnard, California, ~2008. https://www.weather.gov/media/wrh/online_publications/talite/talite0905.pdf.

Kasper, Donald T. "Santa Ana Windflow in the Newhall Pass as Determined by an Analysis of Tree Deformation." *Journal of Applied Meteorology and Climatology* 20, no. 11 (1981): 1267–76. https://journals.ametsoc.org/view/journals/apme/20/11/1520-0450_1981_020_1267_sawitn_2_0_co_2.xml.

McClung, Brandon, and Clifford F. Mass. "The Strong, Dry Winds of Central and Northern California: Climatology and Synoptic Evolution." *Weather and Forecasting* 35, no. 5 (2020): 2163–178. https://journals.ametsoc.org/view/journals/wefo/35/5/wafD190221.xml.

NASA. "How Do We Know Climate Change Is Real?" *Global Climate Change: Vital Signs of the Planet*. Accessed June 15, 2023. https://climate.nasa.gov/evidence/.

NASA Earth Observatory. "California Temperatures on the Rise." Image of the Day for April 17, 2007. https://earthobservatory.nasa.gov/images/7596/california-temperatures-on-the-rise.

NOAA. "What Is El Niño?" https://www.noaa.gov/understanding-el-nino.

NOAA National Ocean Service. "What Are El Niño and La Niña?" NOAA.gov. Accessed June 15, 2023. https://oceanservice.noaa.gov/facts/ninonina.html.

Parkman, E. Breck. "Creating Thunder: The Western Rain-Making Process." *Journal of California and Great Basin Anthropology* 15, no. 1 (1993): 90–110. https://escholarship.org/content/qt7fk1p7v0/qt7fk1p7v0.pdf?t=krnin1.

Reynolds, David W. "Association of Wet and Dry Periods in Northern California and SSTs in the Pacific Ocean." National Weather Service, Monterey, CA, 2010. https://www.weather.gov/media/wrh/online_publications/TAs/TA1004.pdf.

Ripple, William J., Christopher Wolf, Jillian W. Gregg, Johan Rockström, Thomas M. Newsome, Beverly E. Law, Luiz Marques, Timothy M. Lenton, Chi Xu, Saleemul Huq, Leon Simons, Sir David Anthony King. "The 2023 State of the Climate Report: Entering Uncharted Territory." *BioScience*, biad080 (2023). https://doi.org/10.1093/biosci/biad080.

Sanger, Gary E., and James R. Andersen Jr. "Tornado Statistics for the WFO San Joaquin Valley–Hanford County Warning Area." NOAA/National Weather Service, San Joaquin Valley–Hanford, CA, 2014. https://www.weather.gov/media/wrh/online_publications/talite/talite1404.pdf.

State of California. "California's Fourth Climate Change Assessment: California's Changing Climate 2018." 2019. www.climateassessment.ca.gov.

Tornado Archive. "A History of Twisters: Tornadoes in California since 1950." Accessed June 15, 2023. https://data.redding.com/tornado-archive/.

US Environmental Protection Agency. "Pacific Southwest Region 9 Air Quality Analysis: Air Quality Trends." *EPA.gov*. Accessed June 15, 2023. https://www3.epa.gov/region9/air/trends/index.html.

van Wagtendonk, Jan W., and Daniel R. Cayan. "Temporal and Spatial Distribution of Lightning Strikes in California in Relation to Large-Scale Weather Patterns." *Fire Ecology* 4 (2008): 34–56. https://doi.org/10.4996/fireecology.0401034.

Williams, A. Park, Benjamin I. Cook, and Jason E. Smerdon. "Rapid Intensification of the Emerging Southwestern North American Megadrought in 2020–2021." *Nature Climate Change* 12 (2022): 232–34. https://doi.org/10.1038/s41558-022-01290-z.

Zhong, Shiyuan, Ju Li, Craig B. Clements, Stephan F. J. De Wekker, and Xindi Bian. "Forcing Mechanisms for Washoe Zephyr—A Daytime Downslope Wind System in the Lee of the Sierra Nevada." *Journal of Applied Meteorology and Climatology* 47, no. 1 (2008): 339–50. https://journals.ametsoc.org/view/journals/apme/47/1/2007jamc1576.1.xml.

ABOUT THE AUTHOR

William A. Selby is an earth science researcher and teacher. A former professor at Santa Monica College, where he taught for three decades, Selby is the author of the popular textbook *Rediscovering the Golden State: California Geography,* whose fourth edition was published in 2019. He has participated in research and seminars with the National Weather Service, and he continues to present at professional conferences and lead teacher trainings and docent workshops. His academic and practical expertise within California's myriad landscapes make him an invaluable guide to the developments that are changing California's climate in the twenty-first century. To learn about new developments in California weather and geography, follow Selby online at www.rediscoveringthegoldenstate.com.